Herausgeber:

Prof. Dr. *A. Davison*   Department of Chemistry, Massachusetts Institute
of Technology, Cambridge, MA 02139, USA

Prof. Dr. *M. J. S. Dewar*   Department of Chemistry, The University of Texas
Austin, TX 78712, USA

Prof. Dr. *K. Hafner*   Institut für Organische Chemie der TH
6100 Darmstadt, Schloßgartenstraße 2

Prof. Dr. *E. Heilbronner*   Physikalisch-Chemisches Institut der Universität
CH-4000 Basel, Klingelbergstraße 80

Prof. Dr. *U. Hofmann*   Institut für Anorganische Chemie der Universität
6900 Heidelberg 1, Tiergartenstraße

Prof. Dr. *K. Niedenzu*   University of Kentucky, College of Arts and Sciences
Department of Chemistry, Lexington, KY 40506, USA

Prof. Dr. *Kl. Schäfer*   Institut für Physikalische Chemie der Universität
6900 Heidelberg 1, Tiergartenstraße

Prof. Dr. *G. Wittig*   Institut für Organische Chemie der Universität
6900 Heidelberg 1, Tiergartenstraße

Schriftleitung:

Dipl.-Chem. *F. Boschke*   Springer-Verlag, 6900 Heidelberg 1, Postfach 1780

Springer-Verlag   6900 Heidelberg 1 · Postfach 1780
Telefon (06221) 49101 · Telex 04-61723
1000 Berlin 33 · Heidelberger Platz 3
Telefon (0311) 822001 · Telex 01-83319

Springer-Verlag   New York, NY 10010 · 175, Fifth Avenue
New York Inc.   Telefon 673-2660

Fortschritte der chemischen Forschung
Topics in Current Chemistry

Band 16, Heft 2, Januar 1971

NMR Spectroscopy of Annulenes

Springer-Verlag
Berlin Heidelberg GmbH

ISBN 978-3-540-05299-9   ISBN 978-3-540-36436-8 (eBook)

DOI 10.1007/978-3-540-36436-8

Library of Congress Catalog Card Number 51-5497.

Contents

Nuclear Magnetic Resonance Spectroscopy of Annulenes

R. C. Haddon, Virginia R. Haddon, and L. M. Jackman

Chemistry Department, The Pennsylvania State University, University Park, PA 16802, USA

Contents

A. Introduction 105
    1. Hückel's Rule 105
    2. Scope of the Review 106

B. Bond Alternation in Annulenes 109

C. Criteria of Aromaticity 112

D. Theory of the Magnetic Properties of Annulenes 115
    1. π-Electron Diamagnetism 115
    2. Proton Chemical Shifts 124

E. N.M.R. Spectra of the Annulenes 130

F. N.M.R. Spectra of the Quasi-Annulenes 159
    1. Bridged Annulenes 161
    2. Charged Annulenes 190
    3. Homoaromatic Systems 201

G. Addendum 208

H. References 214

A. Introduction

1. Hückel's Rule

The annulenes are that series of monocyclic polyolefins (C_nH_n) containing a complete system of contiguous double bonds. While benzene (the best known member of this class of compounds) has been in evidence for some time it is only of late that interest in the higher members has become apparent. This interest has its origins in the LCAO-MO theory of π-electron systems as formulated by E. Hückel [1] (in particular the "Hückel rule" relating aromatic stability to structure). Although the non-classical chemistry of the benzenoid hydrocarbons had previously been the subject of some conjecture, Hückel's theoretical studies provided the first satisfactory explanation of the peculiar stability of this class of compounds and, incidently, the elusiveness of cyclobutadiene.

Hückel's rule (in its original form) stated that monocyclic polyenic molecules are aromatic only if their π-systems contain $(4n + 2)$ π-electrons, where n is an integer [1]. There have been many advances in LCAO-MO theory since Hückel's original contributions (although the simplest approximation still bears his name, i.e. HMO), and today a more precise statement of the rule might read as follows.

For polyenes with $4n + 2$ carbon atoms in the path of conjugation, the π-electron energies of the annulenes, $C_{(4n+2)}H_{(4n+2)}$ will be lower (i.e. more stable — positive resonance energy[a]) than those of the linear polyenes $C_{(4n+2)}H_{(4n+4)}$, whereas the reverse is true for the annulenes, $C_{4n}H_{4n}$, which have π-electron energies greater than those of the corresponding linear polyenes, $C_{4n}H_{(4n+2)}$ (i.e. less stable — negative resonance energy). These predictions lead to the definition of the concepts

---

[a] Following Dewar [2,3], we depart here from the usual practice [4] of defining the resonance energy of a polyene with respect to the π-energy of the appropriate number of *non-interacting* double bonds (when defined in this manner resonance energies of all annulenes are positive, except that of cyclobutadiene which is zero [2e]). It should be noted however, that in HMO theory the assumption of equal bond lengths is implicit; this is known to be incorrect for linear polyenes which are subject to strong bond alternation. To this extent the definition we adopt here is artificial, but is nevertheless adequate for our purposes.

aromatic and *anti-aromatic* in terms of positive and negative resonance energies, respectively[b]. Hückel's assertions are on strongest ground when applied to neutral alternant hydrocarbons [2] of which the annulenes form the important subgroup of monocyclic systems. Later theories have challenged much of the detail in Hückel's analysis [2a], but the predictions listed above remain substantially unchanged (provided of course, that molecular geometry permits approximate planarity of the π-system). For instance, it is now realised that the [4n]annulenes will be subject to a Jahn-Teller distortion which leads to bond length alternation [5,6]. This lifts the degeneracy of the non-bonding orbitals, and these systems are no longer expected to have triplet ground states (Fig. 1). Furthermore, the resonance energies of both the [4n]- and [4n + 2]-annulenes ultimately converge to zero (non-aromatic behaviour) at large n (and, as a result, bond alternation is also expected for the higher [4n + 2]annulenes).

2. Scope of the Review

A discussion of n.m.r. characteristics is reported for the annulenes (and dehydro analogs), the related anions and cations, various bridged systems, and certain annulene-related systems. The annulenes and their ions are assumed to be cyclic, essentially planar systems involving a single loop of conjugated carbon atoms. The remaining systems contain atoms or groups which may enter into conjugation with the annulene or hetero-annulene ring (e.g. porphine) but do not perturb too seriously the aromatic (or anti-aromatic) nature of the system. Cross-conjugated molecules (e.g. tropone) will not be discussed but homo-conjugated systems are included.

Prior to this discussion of experimental n.m.r. data, a brief theoretical outline of annulene chemistry will be included, which begins with a consideration of the phenomenon of bond alternation in the next section. An understanding of bond alternation is necessary as it augments Hückel's rule in delineating more precisely the relationship between the annulenes and the corresponding linear polyenes. A section will then be devoted to a consideration of some of the more important experimental parameters (other than magnetic properties) which have been studied in relation to aromaticity.

---

[b] For ions the rule leads to the following predictions

$C_{(4n+1)}H_{(4n+1)}$: anion, $(4n + 2)\pi$-electrons. Aromatic
cation, $(4n)\pi$-electrons. Anti-aromatic

$C_{(4n+3)}H_{(4n+3)}$: anion, $(4n)\pi$-electrons. Anti-aromatic
cation, $(4n + 2)\pi$-electrons. Aromatic

Geometry Assumed	Planar Symmetric					Planar Distorted	
Compound	Ethylene C_2H_2	Cyclobutadiene[a] C_4H_4	Benzene C_6H_6	Cyclooctatetraene[a] C_8H_8	Cyclodecapentaene[a] $C_{10}H_{10}$	Cyclobutadiene[a] C_4H_4	Cyclooctatetraene[a] C_8H_8
E[b] (orbital energy diagrams from -2β to $+2\beta$)							
HMO π-electron Energy	$2\,\beta$	$4\,\beta$	$8\,\beta$	$9.66\,\beta$	$12.94\,\beta$	$4\,\beta_2$	$4(\beta_2 + \sqrt{\beta_1^2 + \beta_2^2})$
Resonance Energy I[c]	0	0	$2\,\beta$	$1.66\,\beta$	$2.94\,\beta$		
Resonance Energy II[d,e]	0	$-0.47\,\beta$	$1.02\,\beta$	$0.14\,\beta$	$0.89\,\beta$		

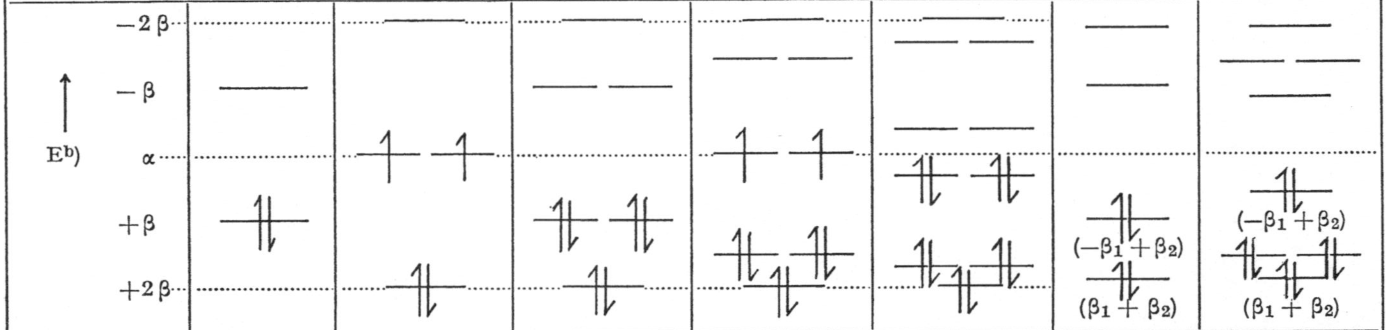

a) Hypothetical cases.
b) β, the carbon-carbon resonance integral, is negative.
c) Defined against appropriate number of non-interacting double bonds [4].
d) Defined against corresponding linear polyene (with bond equalization assumed) [2,3].
e) See also Table 1, and Fig. 2 for SCF MO values [2c].

Fig. 1. Hückel π-electron molecular orbitals for some lower annulenes

The two following sections develop the theory of molecular diamagnetism and proton chemical shifts in aromatic systems thus setting the stage for the detailed review of the n.m.r. spectra of the currently known compounds within our purview. Particular emphasis will be placed on the qualitative conclusions (which may be drawn from the n.m.r. parameters) regarding aromaticity and anti-aromaticity.

B. Bond Alternation in Annulenes

A generally recognized simplification in the HMO theory of polyenes is the assumption of total symmetry (and planarity) in the molecule under consideration[c]. Apart from the geometrical and steric constraints which, in annulenes, cause deviations from planarity and ideal geometry, a more subtle effect (independent of such considerations) is operative. This is the occurrence of bond alternation (which was briefly mentioned earlier).

In the HMO picture the [4n]annulenes possess a pair of degenerate non-bonding orbitals to which two electrons are assigned. The prediction of a degenerate ground state for these molecules is an artifact of the HMO neglect of electron spin and repulsion. On application of self-consistent field molecular orbital techniques (SCF-MO), where allowance for such effects is specifically included, the prediction of orbital degeneracy is retained (by symmetry), but the triplet configuration is no longer degenerate with the three singlet configurations (two of which remain degenerate). Nevertheless the first excited singlet configuration does remain very close in energy to the triplet ground state configuration ("near degeneracy"). Where the ground state of a cyclic molecule is configurationally degenerate, the Jahn-Teller theorem predicts that the molecule will distort so as to remove the degeneracy[d]. The distortion which occurs in the absence of a configurationally degenerate ground state but as a result of a very low-lying excited electronic state (nearly degenerate ground state) is referred to as a *pseudo* Jahn-Teller distortion. It is thus the pseudo Jahn-Teller distortion which is responsible for the bond length alternation observed in the [4n]annulenes. When this bond alternation is incorporated into the MO theory, and the geometry is relaxed from symmetric to distorted, a singlet ground state results (see Figs. 1 and 2, Table 1, and refs. [2,3,6]).

---

[c] Hence in the case of annulene geometry, regular polygons are assumed in the construction of the topological matrix, and the annulene C_nH_n is assigned D_{nh} symmetry.

[d] An example of this is provided by the cyclopentadienyl radical, which has an odd number of electrons (3) to be placed in a doubly degenerate pair of orbitals [6].

So far this discussion has been restricted to the [4n]annulenes where bond alternation is expected for all members, and at first sight the preceding remarks do not appear relevant to the closed-shell [4n + 2]annulenes. However, in 1960, Longuet-Higgins and Salem [7] showed that, within the HMO theory, bond alternation could be expected for the higher [4n + 2]annulenes (with the critical value of n expected to lie between 4 and 8). This idea had its origins in the work of Kuhn [8] and Dewar [9] who investigated the electronic spectra of polyenes and found that the observed frequencies could only be interpreted if the bonds were assumed to alternate (see also ref. [6]). Although the Hückel method as applied by Longuet-Higgins and Salem [7] considered the energy of the ground state as a function of a distortion parameter, alternative approaches explain the destabilization of the bond equalized configuration by focussing on the interaction of the ground state with a low-lying excited state of the correct symmetry (mixing of the ground and excited configurations is spin and/or symmetry forbidden in the bond equalized geometry [6]). In a [4n + 2]annulene as n increasases, the bonding and antibonding levels become closer (i.e. the electronic excitation energies become less). As a result, the mixing of ground and excited states becomes energetically more favourable until for sufficently large values of· n the gain in energy will be such that a distorted configuration (geometry) becomes the ground state. Hence the distortion of the symmetric configurations of [4n]annulenes and higher [4n + 2]annulenes arises from a pseudo Jahn-Teller effect. The result in both cases is a finite energy gap between the filled and unfilled π-orbitals for all values of n. The two classes of annulenes differ in that bond alternation is more pronounced for the lower [4n]annulenes but only becomes important for higher members of the (4n + 2) series. Furthermore the magnitude of the energy gain accompanying distortion is much larger for the [4n]annulenes. However, with increasing n the two series converge to a common limit (corresponding to zero resonance energy and non-aromatic properties).

More recently, Dewar and Gleicher [2c] have carried out SCF calculations for the annulenes (with experimental geometries). From the calculated resonance energies they conclude that the onset of bond alternation in the (4n + 2) series should begin with [26]annulene, which they predict to be non-aromatic (see Table 1 and Fig. 2). Note, however, that the calculated difference in resonance energy between [22] and [26]annulene only amounts to about 7 kcal/mole; in such large molecules this could easily be overshadowed by steric requirements.

Although the available experimental and theoretical evidence in favour of bond alternation for the higher (4n + 2) systems is fairly compelling, a dissonant note is sounded in the alternant molecular orbital calculations of Hultgren, who predicts the total absence of bond alternation [10].

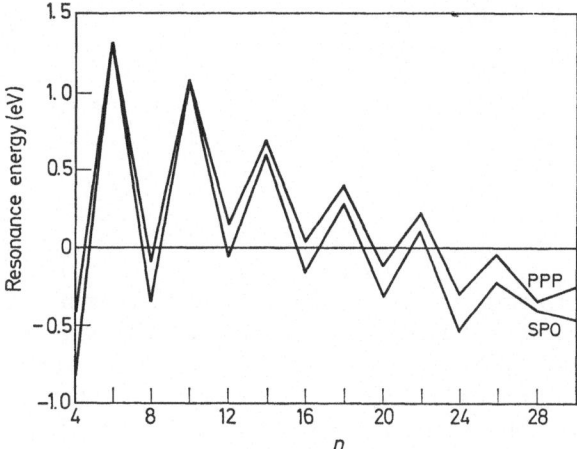

Fig. 2. Calculated[a] resonance energies as a function of annulene ring size (C_nH_n)[2c]. Reprinted with permission of the copyright owner from J. Am. Chem. Soc. *87*, 685 (1965)

[a] PPP refers to the Pariser-Parr-Pople method [15]; SPO refers to the Split *p*-orbital method [2a]; see ref. [2].

Table 1. *Calculated*[a] *π-binding energies and resonance energies of the annulenes* C_nH_n [2c]

n	PPP		SPO	
	$E_{\pi b}$, eV	E_R, eV	$E_{\pi b}$, eV	E_R, eV
4	3.815	—0.534	3.091	—0.815
6	7.841	1.318	7.177	1.318
8	8.589	—0.108	7.485	—0.327
10	11.933	1.061	10.844	1.079
12	13.177	0.131	11.667	—0.051
14	15.909	0.689	14.294	0.623
16	17.438	0.043	15.489	—0.135
18	19.951	0.382	17.854	0.277
20	21.623	—0.120	19.215	—0.315
22	24.147	0.229	21.573	0.090
24	25.785	—0.305	22.901	—0.535
26	28.211	—0.054	25.160	—0.229
28	30.106	—0.339	26.977	—0.365
30	32.350	—0.263	28.824	—0.471

[a] PPP refers to the Pariser-Parr-Pople method [15]; SPO refers to the Split *p*-Orbital method [2a]; see ref. [2]. Reprinted with permission of the copyright owner from J. Am. Chem. Soc. *87*, 685 (1965).

C. Criteria of Aromaticity

Until quite recently the chemical behaviour of benzene was used as a standard by means of which the degree of aromatic character of any compound could (in principle) be established. Particular attention was paid to those reactions in which the structural integrity of the benzene nucleus was preserved but which brought about an irreversible destruction of "normal" olefinic bonds. Apart from the practical difficulties of effecting such a comparison, it is now agreed that a criterion which can be applied to ground state properties rather than transition state energetics is more appropriate to the problem. An obvious candidate is the resonance energy of the system.

It must be remembered that HMO theory assumes that relative stabilities of annulenes are solely determined by π-electron energies and specifically excludes any consideration of the σ-electrons (irrespective of the topological constraints of the hydrocarbon) on the grounds that these skeletal electrons contribute a constant and therefore dispensible term to the total energy. This approximation forms the basis of the principle of σ-π separability [5] which may be used with reasonable impunity in the absence of angle strain and large non-bonded interactions. In many annulenes, however, planarity may only be achieved at the expense of severe non-bonded interactions between interior hydrogens, and as a result the geometrical constraints accompanying cyclization sometimes prove overwhelming. A good example is provided by the (hypothetical) di-*trans*-[10]annulene, *1*. The large non-bonded interac-

1   *2*

tion expected for the interior hydrogens presumably negates the resonance stabilization which HMO theory predicts for a *planar* system. In contrast, 1,6-methano-[10]annulene, *2*, in which a methylene bridge replaces the interacting hydrogen atoms, is known to be a stable molecule [11]. It should thus be clear that in many cases the total energy or relative stability of a particular annulene need not necessarily relate directly to the π-electron energy of the (hypothetical) planar, strain-free compound considered in HMO theory. It must therefore be concluded that heats of formation and hydrogenation are of dubious value in assessing the aromatic character of many annulenes.

It is primarily these considerations which have made it necessary to re-think very carefully what is meant by the aromatic character of a compound in the present day context [12]. Once it has been accepted that the chemical and thermodynamic stabilities of the annulenes do not have very much to do with Hückel's proposals, we may legitimately ask whether benzene might not be the only aromatic annulene. The somewhat unique properties of benzene (particularly with regard to the high resonance energy per electron) have set a rather unattainable standard for aromatic compounds. Untroubled by steric interference, possessed of undeformed bond angles which lead to natural planarity, and of a symmetry such that all carbon atoms are indistinguishable, questions of bond alternation do not arise and benzene may justly be called the perfect annulene. Nevertheless, HMO theory does permit calculation of π-electron energies which are subsequently used in making predictions regarding aromatic character in the annulenes, the significance of which does not appear open to direct experimental verification.

We next consider the use of ultraviolet and visible spectroscopy as this technique has been widely applied in the study of aromatic compounds, particulary the polycyclic benzenoid hydrocarbons [13]. The success of the simple HMO method for predicting ground state properties of π-electron systems is in part due to a fortuitous cancellation of errors inherent in the method [2a, 2e], a situation which does not persist when the method is extended to the calculation of excitation energies. In fact, the work of Goeppert-Mayer and Sklar [14] showed that the observed long wavelength transitions in benzene are not associated with the lowest unoccupied level in the one electron HMO approximation. Indeed, much more sophisticated techniques employing SCF-MO methods [15] and explicit inclusion of configuration interaction are required for a proper understanding of the spectra of aromatic hydrocarbons [16]. As a result of this, and the fact that transition energies are non-ground state properties, u. v. and visible spectroscopy is not a suitable criterion of aromaticity and anti-aromaticity. Nevertheless, the technique has proved useful for the characterisation of iso-structural annulenes, and it is believed that the appearance of three main absorption bands with the longest wavelength transition having the lowest intensity and considerable vibrational fine structure is characteristic of aromatic systems (e.g. see ref. [17]). Furthermore u. v. and visible spectroscopy has played a central role in the investigation of bond alternation in polyenes.

If a quantitative criterion of aromaticity in the annulenes is considered meaningful it is probably best defined in terms of an observed degree of bond alternation, which, as was seen in the preceding section should be related to the resonance energy. Furthermore, spectroscopic, magnetic and diffraction techniques have demonstrated their ability to focus on such a parameter [6]. Perhaps the most powerful method is the direct determination of molecular geometry by single crystal X-ray diffraction or electron diffraction experiments. The completely delocalized nature of benzene follows immediately from the demonstration that it is a perfect hexagon. Similarly, the non-planar and hence non-aromatic nature of cyclooctatetraene was first demonstrated by electron diffraction experiments [18]. For the higher annulenes, the applicability of such a criterion is contingent on the feasibility of single crystal X-ray diffraction studies but unfortunately, few annulenes have been investigated in this manner. Furthermore, the structure of [18]annulene shows the existence of a variation in bond lengths which cannot be classified as bond alternation. These variations are no doubt due to steric factors but what they mean in terms of the aromaticity of this molecule is uncertain. Obviously, however, a determination of bond lengths should unequivocally demonstrate true bond alternation where it exists.

In the next section we take up a discussion of the magnetic properties of the annulenes. In striking analogy to Hückel's rule, an opposite directionality of these properties is found for the [4n]- and [4n + 2]-annulenes. For this reason magnetic properties have played the major role in the investigation of aromatic, non-aromatic and anti-aromatic character in the annulenes. By default, n.m.r. has been the standard experimental probe of annulene magnetism.

D. Theory of the Magnetic Properties of Annulenes

1. π-Electron Diamagnetism

Consider an assembly of non-interacting molecules in which all the electrons are paired (closed shell). In a uniform magnetic field, the motions of the electrons within each molecule are modified so that they flow in a manner corresponding to a circulation of current (Larmor precession). These circulations are usually localized to the individual atoms of the molecule[e]. The currents flow in such a direction that the induced magnetic moments oppose the applied field (Lenz's law), thus giving rise to the phenomenon of diamagnetism [19-22].

In spherically symmetric systems the induced diamagnetism depends primarily on the mean square radius of the valence electrons as the small contribution from the inner-shell electron core can usually be neglected [19]. In the case of molecules with symmetry lower than cubic, the quantum mechanical treatment by Van Vleck [23] indicates that another term must be added to the Larmor-Langevin expression in order to calculate correctly diamagnetic susceptibilities. This second term arises because the electrons now suffer a resistance to precession in certain directions due to the deviations of the atomic potential from centric symmetry. The induced moment will now be dependent on the orientation of the molecule in the applied magnetic field and thus in general the diamagnetic susceptibility will not be an isotropic quantity [19-23].

Pauling Theory

It has long been recognized that benzenoid hydrocarbons are characterised by considerable anisotropies of their large diamagnetic susceptibilities [24]. The first attempt to rationalise the magnetic susceptibility data of such compounds was made by Pauling in 1936, "on the assumption that the $2p_z$ electrons are free to move from carbon atom to ad-

---

[e] This behaviour forms the basis for Pascal's constitutive atom constants from which molecular magnetic susceptibilities may be obtained [19,20].

jacent carbon atom, under the influence of the impressed (magnetic) field" [25,26]. As pointed out by Pauling "A qualitative explanation of these abnormally large susceptibilities as arising from the Larmor precession of electrons in orbits including many nuclei has come to be generally accepted" [24]. This then represents the quantitation of an extremely infectious idea: that in the case of an annulene, the presence of a magnetic field causes a Larmor precession of the $2p\pi$ electrons around the carbocyclic framework — which phenomenon has come to be known as an induced *ring current*. The theory owes much of its attraction to the interpretational simplicity of the model (an electrical current loop). The Larmor frequency of the π-electrons is calculated by using the average component of the magnetic field perpendicular to the plane of the ring, which in conjunction with the number of π-electrons gives the current. Once the current is known the molecular magnetic moment and hence the diamagnetic susceptibility may be determined. As this electronic circulation can only occur in a direction parallel to the plane of the ring it is supposed that there is no in-plane contribution to the susceptibility and in addition it is often assumed that the ring current is the sole source of diamagnetic anisotropy in these systems, with the other diamagnetic contributions expected to be small and isotropic [f]. Pauling's treatment (free electron theory) gave results which were in reasonable agreement with the experimentally determined diamagnetic anisotropy of a number of benzenoid hydrocarbons, but it is now recognised that the model has some serious deficiencies. Although the theory predicts a diamagnetic ring current for any planar conjugated cycle, experimental determinations [28] and subsequent theoretical treatments [29–32] indicate the occurrence of paramagnetism for certain annulenes. It is also worth noting that Pauling's analysis of the currents in the cata-condensed benzenoid hydrocarbons in terms of the induced e.m.f. in an electrical network led to gross overestimates of the diamagnetic susceptibility in higher members. It is generally considered safer to assign each hexagonal cycle an individual magnetic moment equal to that calculated for benzene [33,34].

Like many other quasi-classical methods applied within the domain of quantum mechanics, the free electron theory explained the gross features of the phenomenon and provided an attractively simple physical picture. The method did not of course, suggest any parallels with the quantum mechanically based HMO theory (which predicted an alternation in π-electron properties among the annulenes).

---

[f] Ironically, Pauling who was one of the few early authors to consider additional sources of magnetic anisotropy in aromatic hydrocarbons, was soundly criticised for adding these other contributions into his calculated anisotropies [27a].

London Theory

In 1937 London developed a quantum mechanical treatment of π-electron anisotropy in aromatic compounds which was based on HMO theory [27]. The presence of an applied magnetic field is introduced into HMO theory by modifying the one-electron Hamiltonian so that the quantum mechanical momentum operator includes the vector potential of the field. Unfortunately the specification of the magnetic field does not uniquely determine the corresponding vector potential, even though the two quantities are related within Maxwell's electromagnetic equations. It will suffice to point out that the vector potential at any particular point in a uniform field depends on the position vector of this point referred to some arbitrary coordinate system for which there is no obvious or unique choice of origin. This is not a trivial point, for, whereas one might expect to be able to proceed with a generalized vector potential and finally arrive at a set of equations from which all arbitrariness has been removed, at present there is no obvious way of performing the calculation in this manner, and in practice one is forced to accept an arbitrary origin (gauge) for the potential. However, if, as is usual, we solve the secular determinant using the Hamiltonian with arbitrary gauge and Hückel atomic orbitals, the best linear combination of atomic orbitals will now depend on our choice of gauge and, although this variability in the wave function does not of course affect any of the properties of the system, gauge invariant orbitals are to be preferred.

In order to remove this difficulty London multiplied the Hückel atomic orbitals by an exponential factor which depended on the local vector potential at the atom under consideration; the function so obtained is appropriately called a *gauge invariant atomic orbital* (GIAO). In the formation of the molecular orbitals let us consider the GIAO under the operation of the Hamiltonian. The exponential factor modifies the Hamiltonian in its action on the (field-independent) Hückel atomic orbital so that the vector potential occurring in the momentum operator is now expressed as the difference of the vector potential at the electron (referred to arbitrary gauge), and at the atom (referred to the same arbitrary gauge); this is merely the vector potential at the electron with the atomic centre of the orbital as gauge. Hence the arbitrariness of the calculation is removed and we have a natural (local) gauge which retains the symmetry of the problem. The same GIAO are obtained more rigorously by a consideration of the invariance properties of the Hamiltonian to certain unitary transformations; this approach emphasises the role of the exponential multiplier for the Hückel atomic orbital as an unobservable phase factor [35].

117

By disregarding the contributions from localised terms it is found that the only elements which are modified in the HMO secular determinant by the presence of a field are the resonance integrals of directly bonded atoms (although further assumptions must be made to obtain this result [36]). When the secular determinant is expanded it is found that coefficients of the various powers of the modified energies occur as constants or as products defining closed cycle(s) [37] (in the case of the mono-cyclic annulenes, one current loop only is possible). The coefficients defining the cycles are dependent on the magnetic flux density across the polygon circumscribed by the peripheral bonds of the polyene. As the bonds of the periphery add vectorially to complete the cycle the direction of the induced current flow is implied in the treatment [32].

The other parameters of interest follow quickly from the modified energies (which may be obtained in closed form for the annulenes). The magnetic moment, which is proportional to the induced current, is given by the negative of the first derivative of the modified energies with respect to the applied field. The π-electron susceptibility is then obtained by taking the derivative of the magnetic moment with respect to the applied field, evaluated at zero field strength. Although it still remains to assign a value to the carbon-carbon resonance integral, the theory clearly predicts a negative contribution to the susceptibility (diamagnetic ring current) for the $[4n+2]$annulenes (and the cata-condensed benzenoid hydrocarbons). London did not, however, choose to comment on the magnetic properties of the $[4n]$annulenes, even though he went on to develop much of the necessary theory (by allowing for the occurrence of bond alternation). London treated the carbon-carbon resonance integral (β) as an empirical parameter and adjusted it so that the calculated contribution of the π-electron ring current to the magnetic susceptibility of benzene was equal to the experimentally determined diamagnetic anisotropy. The value so obtained was used for the other hydrocarbons which were considered and it is now customary to report calculated magnetic anisotropies as ratios to the benzene values.

Other Sources of Diamagnetism

London's method [31] for the calculation of π-electron anisotropies was an improvement on the Pauling approach but when applied to the higher benzenoid hydrocarbons still gave values which were somewhat too large. Subsequently, the more refined *ab initio* calculations [31,39] on benzene have indicated the probable reason for this deviation. These calculations which were designed to obviate the need of an empirical value for the carbon-carbon resonance integral (β), indicated that only a part of the total anisotropy arises from the π-electron ring current (the total anisotropy of planar hydrocarbons, it will be remembered, is taken as the difference of the susceptibility in the direction perpendicular to the ring and the average in-plane susceptibility). In the total anisotropy of benzene the following contributions are thought to be important [32, 34, 40]: the π-electron ring current contribution (which is wholly anisotropic) (56%) [41], the contribution from the localised $2p\pi$ electrons (29%) [42], and the anisotropic contribution of the σ-bonds (9%) [43], the sum of these calculated contributions accounting for about 94% of the experimentally observed anisotropy [44a] (see however ref. [45]).

Paramagnetism

It has been subsequently realised that the London theory sometimes predicts a positive contribution to the magnetic susceptibility [29-32, 45, 46]. Baer, Kuhn and Regel [30] obtained a similar result using the one-dimensional electron gas model (free electron molecular orbital method, FEMO). We have already remarked on the Hückel prediction of degeneracy for the non-bonding levels in the [4n]annulenes (implying a triplet ground state) and the intervention of the Jahn-Teller effect which lifts this degeneracy through bond length alternation in this class of compounds. The incorporation of this feature into any theory of the magnetic properties of the [4n]annulenes is therefore essential. Baer, Kuhn and Regel introduced the effects of bond alternation into the FEMO method by the incorporation of a sinusoidal potential, with maxima at the "single bonds" and minima at the "double bonds" which had the effect of introducing a series of potential barriers to the electronic circulations, the magnitude of which could be varied by adjusting the difference in potential at the two types of bonds. Pople and Untch [29] used the London theory in their consideration of the magnetic properties of the annulenes in which the appearance of bond length alternation necessitates the use of two different carbon-carbon resonance integrals (β_1 and β_2). Happily, the predictions of the two theories we have considered are in pleasing accord, and we can discuss their results simultaneously (see Fig. 3 and Table 2).

From the decomposition of the ring current contributions by individual states (Tables 2a and 2b) it may be seen that the populated orbitals of the annulenes usually make a negative contribution to the susceptibility (diamagnetic ring current), sometimes with the exception of the highest occupied level(s). For the [4n + 2]annulenes the contribution of these upper states becomes more positive with increasing bond alternation; nevertheless in all instances the predicted ring current is overall diamagnetic. In the case of the [4n]annulenes, however, the upper

119

Table 2. *Calculated ring currents*[a]) *for annulenes* C_nH_n
Reprinted with permission of the copyright owner from Z. Naturforsch. 22a, 103 (1967)

Table 2a. *Ring currents*[a]) *for benzene and* [18]*annulene* ($J_k = ring$ *current for* k^{th} *electronic state;* $J_{total} = total$ *ring current*)

$n = 6$	$V_0 = 0$ eV	$V_0 = 1.2$ eV	$V_0 = 2.4$ eV	$V_0 = 3.6$ eV
J_1	—2.00	—1.99	—1.94	—1.87
$J_2 + J_3$	—4.00	—3.60	—2.60	—1.46
J_{total}	—6.00	—5.59	—4.54	—3.33

$n = 18$	$V_0 = 0$ eV	$V_0 = 1.2$ eV	$V_0 = 2.4$ eV	$V_0 = 3.6$ eV
J_1	— 2.00	— 1.99	—1.94	—1.87
$J_2 + J_3$	— 4.00	— 3.97	—3.84	—3.65
$J_4 + J_5$	— 4.00	— 3.93	—3.64	—3.26
$J_6 + J_7$	— 4.00	— 3.60	—2.60	—1.46
$J_8 + J_9$	— 4.00	+ 3.47	+9.40	+9.69
J_{total}	—18.00	—10.02	—2.62	—0.55

Table 2b. *Ring currents*[a]) *for cyclobutadiene and* [16]*annulene (as in Table 2a)*

$n = 4$	$V_0 = 1.2$ eV	$V_0 = 2.4$ eV	$V_0 = 3.6$ eV
J_1	— 1.99	— 1.94	—1.87
J_2	+30.13	+14.32	+9.16
J_{total}	+28.14	+12.38	+7.29

$n = 16$	$V_0 = 1.2$ eV	$V_0 = 2.4$ eV	$V_0 = 3.6$ eV
J_1	— 1.99	— 1.94	—1.87
$J_2 + J_3$	— 3.95	— 3.83	—3.63
$J_4 + J_5$	— 3.88	— 3.53	—3.04
$J_6 + J_7$	— 3.08	— 1.13	+0.63
J_8	+30.13	+14.32	+9.16
J_{total}	+17.23	+ 3.89	+1.25

[a]) In units of $He_0^2/4 \pi m_0 c$, see ref. [30]). V_0 is the height of potential barrier; $V_0 = 0$ in the case of bond equalization; $V_0 = 2.4$ eV (55 kcal/mole) in the case of bond alternation with a bond length difference of 0.14 Å [30]).

Table 2c. *Total ring currents*[a] *for* [4]- *to* [18]-*annulenes*.

n	J_{total} for $V_0 = 0$ eV	J_{total} for $V_0 = 1.2$ eV	J_{total} for $V_0 = 2.4$ eV	J_{total} for $V_0 = 3.6$ eV
4	—	+28.14	+12.38	+7.29
6	— 6.00	— 5.59	— 4.54	—3.33
8	—	+24.26	+ 8.85	+4.24
10	—10.00	— 8.16	— 4.88	—2.40
12	—	+20.92	+ 6.00	+2.30
14	—14.00	— 9.61	— 3.87	—1.25
16	—	+17.23	+ 3.89	+1.25
18	—18.00	—10.02	— 2.62	—0.55

[a] In units of $He_0^2/4\,\pi m_0 c$, see ref. [30]. V_0 is the height of potential barrier; $V_0 = 0$ in the case of bond equalization; $V_0 = 2.4$ eV (55 kcal/mole) in the case of bond alternation with a bond length difference of 0.14 Å [30].

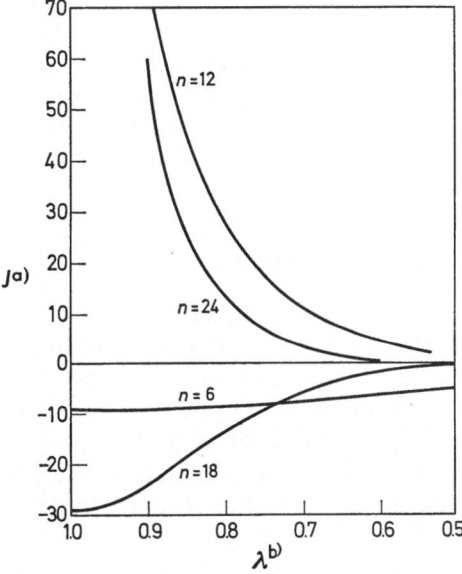

Fig. 3. Ring currents[a] for the annulenes C_nH_n *vs.* the alternation parameter λ[b]. Reprinted with permission of the copyright owner from J. Am. Chem. Soc. *88*, 4811 (1966).

[a] In units of $(\pi^2\,e^2\,\beta_0\,R_{cc}^2\,H/4h^2c)$ see ref. [29].

[b] $\beta_1 = \lambda^{\frac{1}{2}}\beta_0$; $\beta_2 = \lambda^{-\frac{1}{2}}\beta_0$ ($\lambda = \beta_1/\beta_2$; $\lambda \leqslant 1$). β_0 is the value of the carbon-carbon resonance integral in benzene, where $\lambda = 1$ (bond equalization). For $\lambda = 0$ (hypothetical) there would be total bond alternation, with no interaction between adjacent double bonds (this behaviour is approached by cyclooctatetraene, see Table 3).

level always makes a large positive contribution to the susceptibility (which decreases with increased bond alternation) and in all instances overwhelms the diamagnetic contribution made by the remaining states, thus leading to the prediction of a paramagnetic ring current for the [4n]annulenes. For all annulenes, however, the magnitude of the calculated ring current is diminished by the occurrence of the bond alternation and, for a given degree of alternation, the quenching of the ring current is expected to be more effective in larger rings. For the [4n + 2]-annulenes the diamagnetic ring current is predicted to increase with ring size, in the absence of significant bond alternation. However, as the earlier discussion shows, the higher members of the series *are* subject to bond alternation, the onset of which may be further encouraged by the presence of triple bonds and heteroatoms. Finally, we note that distortions from planarity would be expected to reduce the ring current by decreasing the magnitude of $2p\pi$ overlap.

Clearly, experimental data regarding the magnetic anisotropy of the annulenes would be of great interest. The measurements which have been reported [44] and the work on large heterocyclic rings (phthalocyanines [26], etc.) suggest that the predictions we have discussed here might well be borne out in practice [44b]. The considerable difficulties associated with the experimental determination of the magnetic anisotropy of annulenes constitutes the major drawback to such studies.

Diamagnetic Susceptibility Exaltation

In the absence of magnetic anisotropy data, Dauben, Wilson and Laity [28] have resurrected the idea [47,20] that the observed exaltation of diamagnetic susceptibility of a compound might provide a suitable criterion for the presence of an induced ring current, and, as a corollary, the aromatic character thereof. The diamagnetic susceptibility exaltation (Λ) of a compound, is defined as the difference between the experimentally determined molar susceptibility of that compound, and the susceptibility estimated for a model cyclopolyene of that structure (constituted from an increment system [19], without the inclusion of ring current effects [28]). The (mean) magnetic susceptibility (which is considerably easier to determine than the magnetic anisotropy) is equal to the average of the three susceptibility components. Thus the ring current contribution to the mean susceptibility, which is limited to the out-of-plane component, is reduced by one-third. As might be expected, a positive exaltation of the diamagnetic susceptibility is taken to be indicative of a diamagnetic ring current, whereas a negative exaltation is thought to suggest the presence of a paramagnetic ring current. Dauben,

Wilson and Laity claim that the incremental bond system used in the construction of the model [19] will include the Van Vleck paramagnetism of the localised $2p\pi$ electrons. Accordingly, they attribute the observed exaltations to the presence of ring currents (London diamagnetism).

The exaltation observed for benzene assigns 69% of the experimental magnetic anisotropy to the diamagnetic ring current [28]. This high value may be due to the underestimation [32,42] of the local Van Vleck paramagnetism in the construction of the model compound which is necessarily based on strong bond alternation, rather than bond equalisation. Nevertheless, in general, the method appears promising, and it is hoped that more complete data for the annulenes will become available (the currently available values are presented in Table 3).

Table 3. *Diamagnetic susceptibility exaltation data (Λ)* [28]

Compound	X_M [a]	$X_{M'}$ [a]	Λ [a]
Cyclohexane	68.1 [b]	68.1	0.0
Cyclohexene	57.5 [b]	58.3	— 0.8
1,3-Cyclohexadiene	48.6 [b]	49.3	— 0.7
1,4-Cyclohexadiene	48.7 [b]	48.5	0.2
Cyclooctane	91.4 [b]	90.8	0.6
Cyclooctene	80.5 ± 0.6 [c]	81.0	— 0.5
1,3-Cyclooctadiene	72.8 ± 0.8 [c]	72.0	0.8
1,5-Cyclooctadiene	71.5 ± 0.7 [c]	71.2	0.3
1,3,5-Cyclooctatriene	65.1 ± 0.8 [c]	64.0	1.1
Cyclooctatetraene	53.9 [c,d]	54.8	— 0.9
[16]annulene	105 ± 2 [c]	110	— 5
Benzene	54.8 [b]	41.1	13.7
1,6-Methano-[10]annulene	111.9 ± 0.4 [c]	75.1	36.8
1,6-Oxido-[10]annulene	108.0 ± 0.5 [c]	69.1	38.9
trans-15,16-Dimethyl-15,16-dihydropyrene	210 ± 15 [c]	129	81

[a] Λ (diamagnetic susceptibility exaltation parameter) $= X_M - X_{M'}$, all in units of -10^{-6} cm$^3 \cdot$ mol^{-1}. X_M is the experimentally determined molar susceptibility, $X_{M'}$ is the susceptibility estimated for a model cyclopolyene (neglecting ring current corrections) [28].

[b] G. W. Smith, A Compilation of Diamagnetic Susceptibilities, General Motors Corporation Research Report, GMR-317 1960.

[c] Ref. [28].

[d] Shida, S., Fujii, S.: Bull. Chem. Soc. Japan **24**, 173 (1951).

2. Proton Chemical Shifts

Although the experimental difficulties associated with magnetic susceptibility measurements do not carry over into nuclear magnetic resonance (n.m.r.) determinations on the annulenes, much of the methodology and some of the difficulties in interpretation of data are common to both areas. Again, however, the pertinent information is potentially available. From the magnetic susceptibility point of view one is interested in the electronic circulations induced by the magnetic field *per se*, whereas, in n.m.r. we focus on the effects of the induced currents, specifically their associated secondary magnetic fields which we attempt to chart by means of the n.m.r. chemical shift [48,49]. While the presence of spin-spin coupling in the annulenes has been of use in making assignments and estimating bond alternation, primary interest focuses on the proton chemical shifts. Much as with the magnetic anisotropy and diamagnetic susceptibility exaltation work we are again concerned with model compounds and efforts to identify and remove all contributions to the chemical shift other than the contribution arising from the presence of ring currents. Attempts to calculate the ring current contribution to proton chemical shifts have usually been made relative to the observed shift of a polyolefin for which linear but not cyclic delocalisation is possible [50,51]. A fairly representative compound (which has more or less been accepted as a standard), is 1,3-cyclohexadiene the olefinic protons of which resonate at about τ 4.2 ppm [50,51]. This is to be compared with a chemical shift for the protons in benzene of approximately τ 2.7; hence workers in the field have generally assumed that theories of the aromatic chemical shift in benzene must account for a downfield shift (deshielding) of about 1.5 ppm.

Pople Method

In 1956 Pople [52] suggested that the deshielding of the benzene protons may be yet another manifestation of the molecular ring current induced by the presence of a magnetic field. This is a distinct possibility, for, whereas the applied field is opposed by the induced field in the interior of the ring, it is augmented by this secondary field in the neighborhood of the peripheral protons.

In order to estimate the magnitude of this secondary field, Pople replaced the induced current by a point dipole (at the centre of the ring) equivalent to the magnetic moment of the current loop (see Fig. 4). The magnetic moment, it will be recalled, is proportional to the current flowing in the loop (for which Pople used the Pauling [25] free electron value), and the area [34,51] of the conductor (benzene ring). In an isotropic liquid the secondary field at the protons which is estimated from the magnetic dipole must be averaged over all directions as the molecule undergoes rapid reorientation with respect to the applied field. This then gives directly the ring current contribution to the (de)shielding of the protons; surprisingly for such a simple model, the value estimated for benzene, -1.4_5 ppm [51], compares extremely well with the chemical shift which is assumed to accompany aromatisation in this compound (-1.5 ppm).

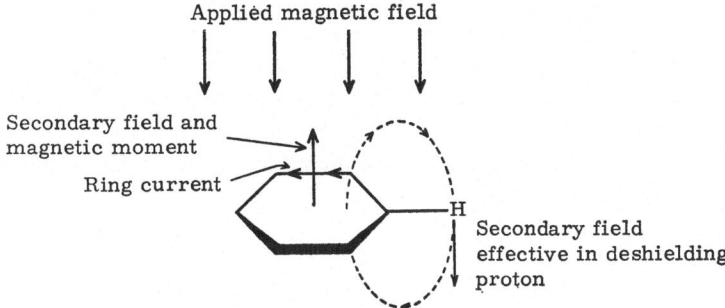

Fig. 4. Secondary magnetic field in benzene

Waugh-Fessenden-Johnson-Bovey Method

Unfortunately, the above-mentioned agreement must be regarded as largely fortuitous in view of the fact that the point dipole method employed by Pople is only valid for points which, in terms of the dimensions of the current loop, are well distant from the centre of the conductor. Clearly, the protons of the benzene molecule do not satisfy this criterion. The procedure used by Waugh and Fessenden [50] improves on Pople's method insofar as the secondary magnetic field arising from the free electron line current in the ring is calculated exactly rather than via an equivalent dipole approximation (although the assumption of circular symmetry in the conductor remains). The magnetic shell obtained in this way predicted the much larger shift of -2.7 ppm for the benzene protons. In an effort to improve the agreement with experiment, Waugh and Fessenden [50] pointed out that the electrons involved in the molecular circulations are formally located in $2p\pi$ orbitals which have a node in the plane of the ring, and proceeded to remove the ring current from the carbon skeleton by splitting it into two current loops symmetrically disposed above and below the hexagon (with radius unchanged). It is difficult to assess the validity of this procedure, particularly as the distance of separation of the two magnetic shells does not have a readily measurable experimental value. In order to obtain agreement between the predicted and "measured" values of the ring current chemical shift in benzene, Waugh and Fessenden set the separation of the magnetic shells at 1.2 Å. Johnson and Bovey [53] presented the results of this method in graphical form, and an extended tabulation is presented in appendix B, volume 1 of reference [48].

While the appeal to classical electromagnetic theory has been somewhat drastic, the methods have proved to be quite successful in practice. These are virtually the only two numerical techniques which find everyday use in the estimation of ring current shielding effects and they have found wide application in the calculation of chemical shifts for protons which are known to be in the vicinity of a benzene ring. For obvious reasons (deviation from circular symmetry, bond alternation, etc.) these two free electron methods are not applicable to the annulenes in general (see however ref. [54]).

Molecular Orbital Methods

Although Pople [38] first applied the London theory of ring current diamagnetism to the calculation of chemical shifts in aromatic molecules, it was McWeeny [55] who showed how the problem could be treated as a unity. Rather than proceeding via the calculation of the ring current, McWeeny included the vector potential due to a magnetic dipole at the proton under consideration (in addition to the vector potential of the applied field) directly into the Hamiltonian of the molecule. By means of this device the total energy and both the magnetic anisotropy and the chemical shift contribution from the ring current can be estimated. McWeeny had to make certain approximations to simplify the calculations and as a result the theory is invalid for points close to the ring [32]. The equations may be solved exactly for simple systems, but the secondary fields calculated are too large (a shift of -2.5 ppm for benzene), comparable in fact to the fields estimated by the uniform magnetic shell method of Waugh and Fessenden [50]. Other molecular orbital methods show the same trends; the SCF [56] and VESCF [57] adaptions which have appeared give estimates of -2.5 and -2.5_8 ppm respectively for the benzene shift (these methods are notable chiefly because they may be applied to ring systems in which unequal charge densities are developed [58]). Because the values are uniformly too large, it has been necessary to scale the results (the factor used is commonly that multiplier (*ca.* 2/3) which gives agreement between the calculated and experimental values for benzene) [56–58]. In obtaining the initial estimates of the benzene chemical shift, the molecular orbital methods use as a standard the experimental diamagnetic anisotropy of benzene in order to estimate certain integrals which are treated as empirical parameters. It will be remembered, however, that the more sophisticated calculations indicated that only a part of the observed magnetic anisotropy of benzene originates from ring current effects; when this is taken into consideration the results become more reasonable.

In 1960 Longuet-Higgins and Salem [7] proposed a simple method for obtaining "exact" chemical shifts from calculated ring currents. Although Longuet-Higgins and Salem use the London method in their calculation of the ring current, any reliable method would suffice. They proceed by considering the line current in each bond of the cycle and, using the Biot-Savart law, obtain the secondary field at a particular proton as a sum of the contributions from each carbon-carbon bond in the molecule. In their original work Longuet-Higgins and Salem calculated the ring current and hence the chemical shifts as a function of the degree of bond alternation in the carbocyclic ring [7]. They applied the method to [18]annulene and obtained remarkably good results (again with reference to the benzene value). With $\lambda = (\beta_1/\beta_2) = 0.8$ (see caption to Fig. 3) the technique predicts the following ring current shifts: outer protons -2.6 ppm, inner protons 8.4 ppm. These results in conjunction with the olefinic chemical shift of τ 4.2 give absolute values of τ 1.6 and 12.6 ppm for the outer and inner protons, respectively (cf. **21** in Table 4). The degree of bond alternation found is not inconsistent with the molecular structure as it is known today (the structure had not been solved at this time, and ideal geometry was assumed). In an independent study Salem [59] applied the method to benzenoid hydrocarbons. For benzene, the *ab initio* SCF ring current predicts a shift of -1.6_7 ppm whereas the London ring current gives a value of -1.0_2 ppm [59]. Although the method of Longuet-Higgins and Salem has largely been ignored, it does appear ideally suited for calculations on the annulenes (which in the main have been lacking; partly due to the difficulties in correctly treating the geometry) and work is currently in progress to test this hypothesis [60].

A word of warning should be inserted here with regard to the model used in all these calculations. The free electron methods treat the ring current as a line current in the carbocyclic skeleton, and it may be argued that the close agreement between the quantum mechanical methods and the (single) circular magnetic shell method with respect to the calculated chemical shift of benzene implies that the same criticism may be levelled at the non-classical methods (and is probably related to the effects of electronic repulsion) [34]. Whether the double magnetic shell method of Waugh and Fessenden [50] (which does attempt to take into account the actual distribution in space of the π-currents) is a valid approach still remains open to question. Even more disturbing, however, is a calculation by Pople [42] which indicates that a part (*ca.* —0.7 ppm) of the aromatic downfield shift in benzene has its origins in local anisotropic effects, and even though the calculation is open to serious criticism [45] it does suggest a re-evaluation of the present procedure which equates the aromatic shift wholly with ring current effects [34,42].

Finally, we should mention the work of Musher [61] who has also discussed the magnetic properties of aromatic compounds, but in a manner which has lead to some controversy. While we feel that there is little excuse for the substitution of dogmatic iconoclasm for reasoned criticism, the fact that the chief protagonists have largely ignored one another's work has hardly served to resolve the scientific questions involved. While no one denies the need for a continuing renaissance in chemical thought one must reflect at length before discarding a theory which has found such wide application and apparent verification. In 1965 Musher published a paper [61a] (and subsequently a review [61b]) which flatly denied the reality of ring currents or any other non-local phenomena and implied that there is a complete analogy between the magnetic susceptibility contributions in aromatics and cycloalkanes. The ring currents arising from London's calculation (and all others) were labelled as artifacts originating from the approximations therein. Not surprisingly this created something of a furor, and when Gaidis and West [62] raised certain experimental objections to this local model, Musher [63] compounded his remarks by suggesting that experimental evidence was irrelevant with regard to his propositions ("anti-theory"). This is a very dangerous statement and always carries by implication a spectre of the countercharge that one's work is not related to physical reality, for it must be remembered that in the final analysis experiment is the arbiter of theory (and not *vice versa*). Musher somewhat moderated his position in a subsequent paper [64] which admits the existence of local and non-local (molecular) terms in the contribution made by the π-electrons to the magnetic susceptibility of benzene. Ironically, Musher also shows that the molecular terms sum to zero in the cycloalkanes, a proof that London himself attempted unsuccessfully in his original paper [27a]. A major criticism of Musher's contribution remains his apparent unwillingness to put numbers to his theory and calculate values for the parameters of interest (for example the magnetic susceptibility of benzene in ref. [64]). Until this is accomplished it will be difficult to assess the power of his method and practitioners in the field will probably adhere to the older theories whose index of reliability is common knowledge.

Conclusion

In summary, we may therefore conclude that while quantitative chemical shift predictions on the larger annulenes have not as yet been reported there is a qualitative agreement on chemical shift directions: protons on the exterior of a $[4n+2]$annulenes will experience a downfield shift while those protons on the interior will suffer an upfield shift of somewhat larger magnitude, both shifts resulting from the presence of an induced diamagnetic ring current [29,30]. The reverse will hold true for $[4n]$annulenes with the exterior protons moving upfield and the interior protons downfield (again to a greater extent) as a result of the induced paramagnetic ring current [29,30]. The presence of any perturbations to the path of conjugation such as non-planarity, severe angle strain (leading to re-hybridisation), the inclusion of triple bonds or hetero-atoms, cross-conjugation or a large degree of bond alternation may be expected to quench, to some extent, the ring current, and, by association, the magnitude of the chemical shift trends detailed above.

We therefore conclude that n.m.r. will differentiate between the $[4n]$- and $[4n+2]$-annulenes, for it is quite apparent that proton chemical shifts are sensitive to the magnetic properties of the π-electrons therein. Insofar as this property parallels the Hückel predictions regarding aromatic and anti-aromatic character (which we now identify with diamagnetic and paramagnetic ring currents) then n.m.r. is a qualitative[g] criterion of aromatic character in the annulenes.

---

[g] The origins of the chemical shifts are probably not sufficiently well understood (as yet), to allow a quantitative discussion of aromatic character in the annulenes. If such a concept is considered meaningful it would probably best be defined in terms of the degree of bond alternation therein, which is of pivotal importance to the π-electron properties (see Sections B and C). Apart from theoretical calculations, a number of physical methods have demonstrated their ability to estimate the extent of bond alternation in annulenes (crystallographic analysis, electronic/vibronic spectral analysis, diamagnetic anisotropy/susceptibility exaltation measurements and of course n.m.r.), see ref. [6] for a full discussion. (Furthermore the known correlation between n.m.r. vicinal coupling constants and carbon-carbon bond orders is of potential utility in any determination of bond alternation [65]).

E. N.M.R. Spectra of the Annulenes

Before embarking on a detailed discussion of the n.m.r. spectroscopy of the annulenes, some general remarks are appropriate. Although Hückel published his findings almost forty years ago, it has only been in the past decade with the pioneer work of Sondheimer and his co-workers that extensive experimental verification of Hückel's predictions has been realized.

In all cases the structures given for the various annulenes have been established utilizing some or all of the following techniques. Hydrogenation to the saturated carbocycle establishes the monocyclic nature of the annulene. The molecular formula is determined by elemental analysis, or more recently and reliably, by accurate mass measurement with a high resolution mass spectrometer. Infrared, ultraviolet spectroscopy, and in some cases X-ray data further help to establish uniquely the structures. The n.m.r. spectrum, by indicating the number and type of protons present, confirms the structure, and, in particular, demonstrates handsomely the aromatic nature of the molecule by reflecting the presence or otherwise of an induced ring current. References to the synthesis and structural elucidation of each compound will be given, and for those cases where a structure is uncertain, the various possibilities will be cited.

For many of these compounds the interpretation of the n.m.r. spectrum is complicated (or aided!) by a temperature dependence which arises because various protons can be interchanged between different environments within the molecule. The mechanism of interchange may be the rotation of a *trans* double bond[h] in the ring, sometimes in conjunction with a valence isomerization.

If such a process is fast on the n.m.r. time scale, each proton will resonate at a τ-value which represents the average of the chemical shifts at each of its environments. This averaged spectrum we will refer to as a "mobile" spectrum [66]. The spectrum may be a simple average, if the

---

[h] The process "rotation of a *trans* double bond" should not be confused with "rotation about a double bond". The former, in fact, involves a co-operative rotation about the two carbon-carbon single bonds adjoining the double bond, which preserves the *trans* stereo-chemistry throughout the process.

interconversion involves equivalent conformers, or a weighted average in cases where the interconverting conformers are unequally populated. On decreasing the temperature of such a sample it is often possible to slow down the proton interchange to such an extent that discrete spectra of each conformer are observed. In such a "non-mobile" spectrum [66] the inner and outer protons of an annulene are more clearly recognisable, and more definite conclusions as to the aromaticity of the molecule in question are possible. The dehydroannulenes are less susceptible to rapid interchange processes than the corresponding annulenes, making the informative non-mobile spectrum more readily attainable.

The smallest conceivable (neutral) annulene is cyclobutadiene, and while it appears fairly certain that this molecule (a $4n\pi$ system, where $n = 1$) occurs as a transient intermediate in certain reactions [67], the parent hydrocarbon has not been isolated. On the basis of the experimental evidence and theoretical predictions to date, the molecule appears to be unstable with respect to dimerization, and little hope remains for its long lived existence.

Benzene, being a strain-free, planar $(4n + 2)\pi$ system $(n = 1)$ is the ideal aromatic compound, and has a reported chemical shift of τ 2.734 [68].

All the protons of cyclooctatetraene ($4n\pi$, with $n = 2$) resonate at a single frequency of τ 4.309 [68]. This magnetic equivalence is brought about by two processes, a ring inversion between two tub-shaped forms, and bond shifts, yielding the two structures $3a$ and $3b$, which, in the absence of substitution are equivalent. Studies of the n.m.r. temperature

$3a$ $3b$

dependence of derivatives of 3 indicate that ring inversion takes place much more rapidly than bond shift, and this inversion probably proceeds via a planar transition state with alternate single and double bonds [69,70] whereas the bond shift probably involves a less favorable, nearplanar transition state with all bond lengths equal [69] (see however, ref. [70]).

[10]Annulene, the next highest analog of benzene in the $(4n + 2)\pi$ series, with $n = 2$, is of considerable theoretical interest. It has proved to be very elusive, however, due no doubt to the unfavourable geometry of the possible isomers.

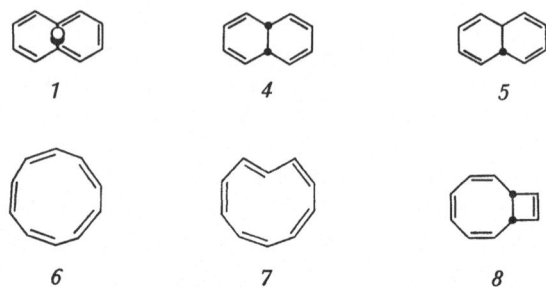

The structure possessing two *trans* double bonds, *1*, can be considered as a valence tautomer of dihydronaphthalene, *4* and *5*. The large non-bonded interaction expected [71] for the internal hydrogen atoms of *1* presumably precludes any approach to planarity. The all-*cis* configuration *6* possesses no such unfavorable interaction, but must involve considerable ring strain, as the inner bond angles in a planar molecule would be 144 °. Again, ring strain is expected for *7* [72]. Despite these unpromising considerations, approaches to the synthesis of cyclodeca-pentaene have been made, with varying degrees of success [i].

Mono-*trans*-[10]annulene, *7*, has been postulated as an unstable intermediate in the conversion of bicyclo[6.2.0]deca-2,4,6,9-tetraene, *8*, to *trans*-9,10-dihydronaphthalene, *5* [76]. It is also claimed that [10]annulene is an intermediate in the photocatalyzed isomerization of *trans*-9,10-dihydronaphthalene, *5*, to the *cis* isomer, *4*, and that it can be trapped at low temperatures by diimide reduction to cyclodecane [77]. These last mentioned results [77], however, have been challenged in a recent communication [72]. Thus, *cis*-9,10-dihydronaphthalene *4*, not the *trans* isomer, is reported to yield [10]annulenes, *6* and *7*, in moderate amounts under carefully controlled conditions. The photolysis of *4* between −50 °C and −70 °C was followed by n.m.r., and the emergence of two new signals was observed. A signal at τ 4.16 (−40 °C) which broadened at lower temperatures was attributed to *7*, and a sharp, temperature-independent singlet at τ 4.34 was assigned to *6*. The structural assignments for these two products are supported by their isomerization to 9,10-dihydronaphthalenes, which was followed by n.m.r.; *7* isomerized to *5* at *ca.* −25 °C, and *6* to *4* at *ca.* −10 °C. Furthermore, hydrogenation (rhodium catalyst) of the compounds responsible for the signals at τ 4.16 and τ 4.34 resulted in the formation of cyclodecane.

---

[i] Annelated derivatives of [10]annulene have been prepared but all of these compounds show n.m.r. spectra typical of non-planar systems, and do not represent delocalized 10π-electron systems [73,74,75].

It is concluded that the two [10]annulenes exhibit no significant dia-magnetic current around the ring system, as their signals appear in a non-aromatic region. However, in the light of conflicting reports, con-firmatory data would be welcome. In particular, the low temperature (down to —80 °C) broadening of the signal of mono-*trans*-[10]annulene should be further investigated, as a similar phenomenon has heralded the approach of non-mobile spectra in the case of [14]annulene, and higher homologues (see 9[j], 10 and 14, Table 4). Should such non-mobile spectra be obtained a more meaningful discussion of the aromaticity of this compound could be entertained.

The [12]annulenes, being $4n\pi$ systems, ($n=3$), are expected to be anti-aromatic. This prediction has been substantiated by the n.m.r. investigation of three compounds in this series.

1,5,9-Tridehydro-[12]annulene [78,66], 1, (see Table 4, p. 134) shows a singlet at τ 5.55 in carbon tetrachloride [78], arising from the six equi-valent outer protons. That the chemical shift of this singlet, which per-sists at least to —80 °C [79], does in fact reflect the presence of a paramag-netic ring current is shown by the comparison of 1 with a model com-pound [80], *cis*-hex-3-ene-1,5-diyne, 9. The olefinic protons of 9 resonate at τ 4.11, indicating that the upfield shift of 1 from 9 arises from a para-magnetic ring current and not soley from the anisotropy of the triple bond [29].

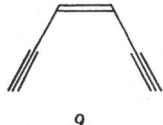

9

1,5-Didehydro-[12]annulene, 2 [81], (see Table 4) possesses one inner proton [H(1')] and seven outer protons, but at room temperature the spectrum indicates that H(1') is magnetically equivalent to the other proton [H(1)] attached to the single *trans* double bond. Furthermore, H(2) is magnetically equivalent to H(2'), H(3) to H(3') and H(4) to H(4'). The most likely explanation of this phenomenon is that the mole-cule is rapidly interconverting between two equivalent conformers, one with H(1) as the inner proton, the other having H(1') as the inner proton [29]. This interchange arises from rotation of the *trans* double bond, a process which is still operative at —80 °C. Such a process has been observed for many of the larger annulenes, and will be described in due course. In the present case a hypothetical non-mobile spectrum can be inferred [29,66] by assuming that, in a frozen conformer, H(1) will have a chemical shift similar to that of the other outer protons.

[j] Bold-faced arabic numerals refer to formulae given in Tables. Italicised arabic numerals pertain to formulae in the running text.

133

Table 4*. *Unbridged annulenes* [g]

Compound	Chemical shifts (τ)	Coupling constants (Hz)	Solvent, temperature	Ref.
1 1,5,9-Tridehydro-[12] annulene	5.55 (s.) (also quoted at 5.58 [66,80])	$J_{H-H} = 10.5 \pm 1$ $J_{13_{C-H}} = 169 \pm 1$	CCl$_4$ Spectrum unchanged to —80 °C See Ref. [79] for other solvents and temperatures	[78]

	Chemical shifts (τ)	Coupling constants (Hz)	Solvent, temperature	Ref.
2 1,5-Didehydro-[12] annulene	Mobile spectrum [a] —0.90 (q., H$_1$, H$_{1'}$) 4.97 (o., H$_2$, H$_{2'}$) 5.47 (H$_4$, H$_{4'}$) 5.82 (H$_3$, H$_{3'}$)	$J_{2,3} \approx 5$ $J_{2,3'} = 2.5$ $J_{1,2} \approx 10.5$	CCl$_4$ Non-mobile spectrum persists to —80 °C	[81,90]
	Non-mobile spectrum		Inferred from R.T. spectrum	[29,82, 66]
	Inner protons	Outer protons		
	< —1.55 (presumably ca. —7)	ca. 4.8—5.8		

---

* The following abbreviations will be used in Tables throughout this text:

s. — singlet
d. — doublet
d.d. — double doublet
t. — triplet
q. — quartet
quin. — quintet
o. — octet
m. — multiplet
b. — broad band
c. — complex
lH — area corresponding to one proton: similarly for 2H, etc.
h.h.w. — width of peak at half height
R.T. — room temperature
THF — tetrahydrofuran
DMSO — dimethylsulphoxide
DME — dimethoxyethane
TFA — trifluoroacetic acid

Table 4 (continued)

Compound	Chemical shifts (τ)	Coupling constants (Hz)	Solvent, temperature	Ref.
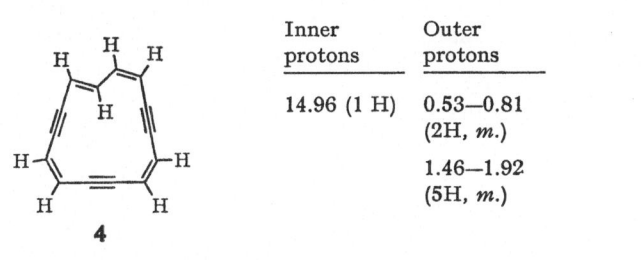 **3** 5-Bromo-1,9-didehydro-[12]annulene	—6.4 (H$_1$) 4.6—5.7 (all other protons)	$J_{1,2} \approx 9.4$ $J_{2,3} \approx 10.2$	CCl$_4$ R.T.	82)

	Inner protons	Outer protons		
4 1,5,9-Tridehydro-[14]-annulene	14.96 (1 H)	0.53—0.81 (2H, *m*.) 1.46—1.92 (5H, *m*.)	CCl$_4$ R.T.	66,85)

	Inner protons	Outer protons			
5 1,8-Didehydro-[14]-annulene	15.48 (*t*., H$_3$)	0.36 (*q*., H$_2$) 1.46 (*d*., H$_1$)	$J_{2,3} = 13.3$ $J_{1,2} = 8.0$	CDCl$_3$ R.T. Spectrum unchanged to 100 °C.	66,87, 90)

Table 4 (continued)

Compound	Chemical shifts (τ)		Coupling constants (Hz)	Solvent, temperature	Ref.
	Inner protons	Outer protons		CCl_4	[66,90]
				R.T.	
	10.7 (2H, *d.d.*)	1.4—2.7 (8H, *c.*)			

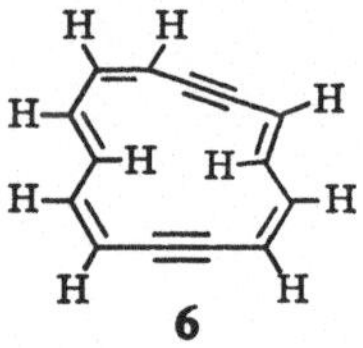

6

1,7-Didehydro-[14]-annulene

	Inner protons	Outer protons	$J_{1,2} = 14.8$	$CDCl_3$ R.T.	[66,90, 91,92]
			$J_{3,2} = 12.3$		
	10.7 (2H, *d.d.*)	1.5—2.7 (10H, *c.*)			

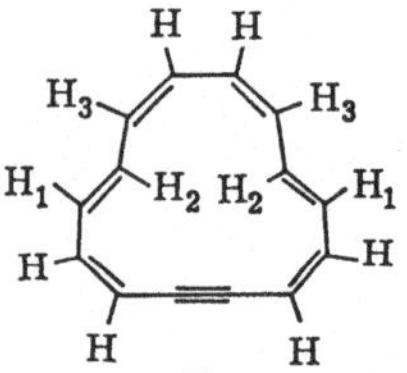

7

Monodehydro-[14]annulene
"Stable conformer" [g]

	Inner protons	Outer protons		$CDCl_3$ R.T.	[66,90]
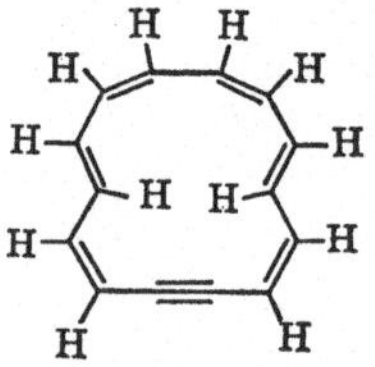	10.6 (2H, *m.*)	1.6—2.6 (10H, *c.*)			

8

Monodehydro-[14]annulene.
"Unstable conformer" [g]

Table 4 (continued)

Compound	Chemical shifts (τ)		Coupling constants (Hz)	Solvent, temperature	Ref.
	Mobile spectrum			CDCl$_3$ *ca.* 40 °C	66,90, 91,93, 95)
	4.42 (*s.*)				
	Non-mobile spectrum				66,90, 93,95)
	Inner protons	Outer protons			
[14]Annulene. "Isomer A". H$_1$ and H$_4$ above H$_8$ and H$_{11}$. [g)	10.0 (4H, h.h.w., *ca.* 25 Hz)	2.4 (10H, h.h.w., *ca.* 40 Hz)		CDCl$_3$ —60 °C	
	Mobile spectrum			CDCl$_3$ *ca.* 40 °C	66,90, 91,93, 95)
	3.93 (*s.*)				
[14]Annulene. "Isomer B". H$_1$ and H$_8$ above H$_4$ and H$_{11}$. [g)					
	Mobile spectrum [b)				98,99, 100)
	0.8 (*c.*, H$_3$, H$_8$)		$J_{7,8}=15$ $J_{1,2}=11$	acetone-d_6 35 °C	
	1.5 (*m.*, H$_4$)		$J_{9,10}=11$		
	2.65 (*d.d.*, H$_7$)		$J_{2,3}=8$ or 9		
	3.95 (*d.d.*, H$_2$ or H$_9$)		$J_{9,8}=9$ or 8		
	4.30 (*d.d.*, H$_9$ or H$_2$)		$J_{5,6}=11$		
	4.55 (*d.d.*, H$_5$)		$J_{5,4}=8.5$		
1,3,9-Tridehydro-[16]-annulene (one of the four possible conformers).	5.0—5.4 (*c.*, H$_1$, H$_6$, H$_{10}$)		$J_{6,7}=2.5$		

9

10

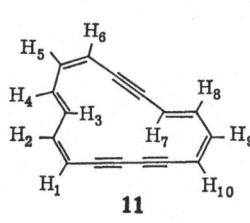

11

Table 4 (continued)

Compound	Chemical shifts (τ)	Coupling constants (Hz)	Solvent, temperature	Ref.
11 (continued)	Non-mobile spectrum[b] —3.60 (0.4H, *d.d.*, H_{3int} or H_{4int}) —3.05 (0.4H, *d.d.*, H_{8int}) —2.00 (0.6H, *d.d.*, H_{3int} or H_{4int}) 1.10 (0.6H, *d.*, H_{7int}) 2.40 (0.6H, *d.d.*, H_{8ext}) 3.5—5.6 (7.5H, *c.*, H_{3ext}, H_{4ext}, H_{7ext})	$J_{7,8} = 15$ $J_{8ext,9} = 5$ $J_{8int,9} = 12$	acetone-d_6 —75 °C	98)

12

1,9-Didehydro-[16]annulene (no conformation implied).

	Mobile spectrum[c] 2.12 (*q.*, H_3, $H_{3'}$) 4.12 (*o.*, H_2, $H_{2'}$) 4.73 (*d.*, H_1, $H_{1'}$)		acetone-d_6 37 °C	66,98)
	Non-mobile spectrum			

Inner protons	Outer protons			
0.2 (2H)	3.92—4.55 (10H)		acetone-d_6 —80 °C	66,98)

13

1,3-Didehydro-[16]annulene (one of the four possible conformers).

	Non-mobile spectrum[b] —3.05 (*d.d.*, H_7) —0.53 (*d.d.*, H_3 or H_{10}) —0.48 (*d.d.*, H_3 or H_{10}) 1.8 (*d d*, H_4) 1.87 (*d.d.*, H_9) 4.42 (*d.d.*, H_2 or H_{11}) 4.55 (*d.d.*, H_2 or H_{11}) ~4.8 (*d.d.*, H_6) ~4.8 (*d.d.*, H_8) 5.25 (*d.d.*, H_5) 5.50 (*d.*, H_1, H_{12})	$J_{7,8} = 15.8$ $J_{7,6} = 11$ $J_{4,3} = 15.5$ $J_{4,5} = 8$ $J_{9,10} = 15.5$ $J_{9,8} = 9$ $J_{5,6} = 11$ $J_{5,4} = 8$	acetone-d_6 0 °C	98)

Table 4 (continued)

Compound	Chemical shifts (τ)		Coupling constants (Hz)	Solvent, temperature	Ref.
	Mobile spectrum			$CS_2/$ CD_2Cl_2	
	3.30 (s.) (also quoted at 3.27 in CCl_4 [90])			30 °C	66,99, 103)
	Non-mobile spectrum				
	Inner protons	Outer protons		$CS_2/$ CD_2Cl_2	103, 104)
	—0.32 ($\sim$4H, t.)	4.8 ($\sim$12H, m.)		—120 °C	

14

[16]Annulene
(One of the two configurations in equilibrium in solution.)

2.98 (s.) — CCl$_4$ — 80)

15

1,3,7,9,13,15-Hexadehydro-[18]annulene.

16

1,3,7,13-Tetradehydro-[18]-annulene (one of the four possible structures).

	Inner protons	Outer protons			
	7.66 (q.) 7.80 (q.)	3.0—3.5 (m., H$_{1a}$, H$_{1b}$) 2.2—3.0 (m., H$_{4a}$, H$_{4b}$, H$_{5a}$, H$_{5b}$) 2.03 (q., H$_{2a}$, H$_{2b}$)	$J_{3,4}=15$ $J_{3,2}=12$ $J_{1,2}=10$	CCl$_4$ 60 °C	107)

Table 4 (continued)

Compound	Chemical shifts (τ)			Coupling constants (Hz)	Solvent, temperature	Ref.

	Inner protons	Outer protons	CH_3	$J_{1,2}=16$	THF—d_8	[108]
	($t.$, H_1)	($d.$, H_2)	($s.$)			
	15.66	0.10	6.65		—60 °C	
	15.24	0.34	7.42		22.5 °C	
	14.99	0.44	7.78		64 °C	

17

1,5,10,14-Tetramethyl-6,8,-15,17-tetradehydro-[18]-annulene.

	Inner protons	Outer protons			
	8.26 ($d.d.$, H_3)	2.98 ($d.d.$, H_1)	$J_{3,4}=15$	CDCl$_3$ 60 °C	[66,90, 91,107]
		2.44 ($d.d.$, H_4)	$J_{2,3}=12.2$	Spectrum esssentially	
		1.90 ($d.d.$, H_2)	$J_{1,2}=10.0$	unchanged at 150 °C.	

18

1,7,13-Tridehydro-[18]-annulene, isomer I.

	Inner protons	Outer protons			
	8.17 (2H, $q.$)	1.7—3.1 (9H, $c.$)	$J_{3,2}=12$	CCl$_4$ 60 °C	[66,90, 107]
	8.28 (1H, $q.$)		$J_{3,4}=15$		

19

1,7,13-Tridehydro-[18]-annulene, isomer II

Table 4 (continued)

Compound	Chemical shifts (τ)		Coupling constants (Hz)	Solvent, temperature	Ref.
	Inner protons	Outer protons			
20	8.4—9.2 (3H, *c.*)	0.64 (*d.*, H_2) 1.6—3.0 7H, *c.*)	$J_{2,3} = 12$	CDCl$_3$	111)
3-Nitro-1,7,13-tridehydro-[18]annulene					

	Mobile spectrum				
	4.55 (*s.*, h.h.w. *ca.* 5 Hz)			THF-d_8 110 °C	66,90, 95)
21	Non-mobile spectrum				
	Inner protons (6H, *quin.*)	Outer protons (12H, *q.*)		THF-d_8	
	12.26	0.97		0 °C	
	12.49	0.89		—20 °C	
	12.75	0.80		—40 °C	
	12.99	0.72		—60 °C	
[18]Annulene	14.22	0.75		toluene-d_8, —60 °C	101)

	Mobile spectrum				
22	1.4—2.0 (4.7H, *b.*) 6.7 (12.3H, *b.*)			toluene-d_8 100 °C	111)
	Non-mobile spectrum				
	Inner protons	Outer protons			
	13.0—14.0 (6H, *c.*)	—0.5—0.7 (11H, *c.*)		THF-d_8 —70 °C	
Nitro-[18]annulene					

Table 4 (continued)

Compound	Chemical shifts (τ)		Coupling constants (Hz)	Solvent, temperature	Ref.

23

Acetyl-[18]annulene

Non-mobile spectrum			CDCl₃	111)

Non-mobile spectrum

Inner protons	Outer protons
12.3—13.3 (6H, c.)	0.0—1.3 (11H, c.)

6.55 (s., CH₃)

Solvent: CDCl₃ —60 °C   Ref. 111)

24

1,11-Didehydro-[20]-annulene

Mobile spectrum[e]

1.60 (m., H₃, H₃′)
2.55 (m., H₄, H₄′)
4.32 (m., H₂, H₂′)
4.91 (m., H₁, H₁′)

THF-d_8
37 °C   66)

Non-mobile spectrum

Inner protons	Outer protons
—1.6 (2H, b., H₃′)	4.40 (c., H₂, H₂′, H₃, H₄′)
—0.45 (2H, b., H₄)	4.93 (c., H₁, H₁′)

THF-d_8
—80 °C

25

1,7,13,19-Tetradehydro-[24]annulene.

Non-mobile spectrum

Inner protons	Outer protons
1.60 (4H, d.d.)	4.40—5.02 (12H, c.)

(See Ref. 91) for shifts in CDCl₃)

$J_{1,2} = 15.5$
$J_{2,3} = 11.0$

toluene-d_8
37 °C
Spectrum essentially unchanged at 150 °C   66,91)

Table 4 (continued)

Compound	Chemical shifts (τ)		Coupling constants (Hz)	Solvent, temperature	Ref.
	Mobile spectrum[f]				
	2.75 (s.)			THF-d_8 40 °C	117)
	Non-mobile Spectrum				
	Inner protons	Outer protons			
	—2.9—1.2 (b.)	5.27 (b.)		THF-d_8 —80 °C	
26	Intensity Ratio 1 : 1.86				
[24]Annulene (one of the two most likely conformations).					
	2.0—4.5 (very broad m.)			CDCl$_3$ or acetone-d_6 R.T. Spectrum unchanged on cooling to —60 °C	118)
27 Tridehydro-[26]annulene (one of many possible structures).					

a) The validity of this analysis has recently been questioned [98]. In acetone-d_6 these chemical shifts occur at τ —1.18, 4.84, 5.32, and 5.73 resp. [66].
b) Assignments based on frequency-swept double-irradiation experiments [98].
c) In CCl$_4$ these chemical shifts occur at τ 2.25, 4.35 and 4.93, resp. [90,92].
d) These assignments based on the assignment of the 15 Hz coupling to the H$_3$, H$_4$ interaction [91].
e) Assignments based on decoupling experiments [66].
f) The chemical shift in CDCl$_3$ at 40 °C is τ 2.78 [117] (c.f. that reported in Ref. [66] and [91], which is erroneous).
g) See also Addendum, p. 208.

143

Thus, the quartet at τ —0.90 [H(1), H(1')] could be considered as the average of a very low field signal at about τ —7, and a higher field signal within the range of τ 4.8 to 5.8. In this system long range shielding by the two triple bonds would augment the contribution of the para-magnetic ring current to the downfield shift of the inner proton(s).

The n.m.r. of a substituted derivative, 5-bromo-1,9-didehydro-[12]annulene [82], 3, (see Table 4) bears out this prediction. This compound can be considered as a prototype for a frozen conformer of 2, because the large bromine atom will strongly prefer the outer position and only one conformer will be significantly populated. Thus, the one proton on the *trans* double bond is held in an inner position, and it is seen to resonate well within the predicted range (τ —6.4). The magnitude and direction of this effect indicates the presence of an induced paramagnetic current in 3, and, by analogy, in the parent compound 2 (the long range shielding effect of the triple bonds still remains an unknown factor). The coupling constant between protons 1 and 2 in 3 (see Table 4) has a value of about 9.4 Hz. In the unsubstituted compound 2 the averaged value of $J_{1,2}$ and $J_{1',2'}$ is 5 Hz. If $J_{1',2'}$ is assigned a value of about 9 Hz by analogy with $J_{1,2}$ in 3, $J_{1,2}$ must then be of the order of 1 Hz. This very low value no doubt reflects a substantial deviation from plan-arity in 1,5-didehydro-[12]annulene. Nevertheless, the mobile, room temperature spectrum of 2 in itself indicates that 2 is anti-aromatic, as the protons fixed in an outer position resonate in a region which is not aromatic, but which is shifted upfield relative to typical cyclic mono-olefins [82].

The parent hydrocarbon of this series, [12]annulene, *10*, is probably still unknown*, a fact which presumably reflects the unfavorable steric interactions [71] of the three internal hydrogens in such a molecule [81, 83,84].

10

The [14]annulenes are within the Hückel criterion for aromaticity ($n = 2$) and a diamagnetic ring current is expected; the spectrum of each compound so far investigated in this series tends to verify such a predic-tion (although a variety of temperature conditions are required to obtain satisfactory non-mobile spectra).

---

* The preparation and an n.m.r. study have now been reported (see Addendum, p. 208).

144

At room temperature 1,5,9-tridehydro-[14]annulene [85,66], **4**, (see Table 4) shows a large shielding of its one inner proton (τ 14.96) and a deshielding of the seven outer protons (τ 0.53 to 1.92).

1,8-Didehydro-[14]annulene [86,87], **5**, (see Table 4) is unusual in that no conventional formula of alternating single and double or triple bonds can be written to represent its structure. Rather it is best represented by the resonance hybrid given in Table 4, which encompasses such cumulene-containing canonical forms as *11a* and *11b*. X-ray analysis [87,88] has

11a   *11b*

indicated that the molecule is planar, despite the close approach of the two internal hydrogens which a planar structure implies. The n.m.r. is typically aromatic; the two inner protons resonate at τ 15.48 (the most shielded of the annulenes) while the outer protons are strongly deshielded (τ 0.36 to 1.46)[k].

A positional isomer of **5**, 1,7-didehydro-[14]annulene [66,90], **6**, (see Table 4) also exhibits chemical shifts characteristic of an aromatic compound, the inner protons (τ 10.7) and the outer protons (τ 1.4 to 2.7) covering a somewhat narrower range (9.3 ppm) than the 15 ppm separation observed for **5**.

Monodehydro-[14]annulene (see Table 4) appears to be isolable in at least two forms, one being called the "stable conformer" [66,86,91,92], **7**, and the other the "unstable conformer" [66,90], **8**, which changes to **7** on standing. Two possible structures for **7**, *12a* and *12b*, have been considered,

12a   *12b*

although the unsymmetrical structure *12b* appears less likely from the preliminary X-ray crystallographic analysis [66]. N.m.r. integration has shown that the both **7** and **8** contain four *cis* and two *trans* double

---

[k] For the n.m.r. spectra of the 3-nitro and the 3-monomethyl-sulphonate of *5*, see Ref. [89].

145

bonds but the structures represented in Table 4 have not been unequivocally established (see however Addendum, p. 209). Both conformers have similar spectra, the two inner protons resonating at about τ 10.7 and the seven outer protons in the range τ 1.5 to 2.7 [1].

When the n.m.r. of [14]annulene [86] was first investigated [91,92] it was seen to consist at 40 °C of two singlets at τ 3.93 and τ 4.42, in a ratio of approximately 1:6. Subsequently it became apparent [93] that each of these singlets represented the mobile spectrum of a distinct isomer of [14]annulene. These isomers differ simply in the spatial position of the four inner protons relative to each other, a phenomenon which is possible because the carbon skeleton is not perfectly planar [94,71]. The two isomers 9 and 10 can be separated by thin layer chromatography on Kieselgel G coated with silver nitrate, but solutions of either isomer rapidly convert to an equilibrium mixture, the composition of which is dependent on temperature [93]. Crystalline [14]annulene [86] appears to consist mainly of "Isomer A", 9, the n.m.r. of fresh solutions showing virtually only a singlet at τ 4.42. That this more plentiful isomer is the one represented as 9 in Table 4 is suggested by the X-ray study at room temperature which indicates a centrosymmetric molecule [94] (allowing for a molecular packing disorder).

When fresh solutions of Isomer A are cooled [95] the spectrum undergoes marked changes. The singlet at τ 4.42 broadens until at -30 °C it can no longer be seen. At -60 °C the presence of two new peaks at τ 10.0 and τ 2.4 represents the non-mobile spectrum, the peaks being assigned to the four inner protons and the ten outer protons, respectively. In compliance with Hückel's rule Isomer A of [14]annulene thus has a non-mobile spectrum typical of an aromatic compound, despite the lack of perfect planarity. The spectrum of Isomer B, however, is unchanged on cooling to -60 °C, except for a slight broadening of the singlet at τ 3.93 [66]. It is to be expected that the non-mobile spectrum could be eventually observed at lower temperatures (for further conclusions concerning these isomers see Addendum, p. 209).

In the [16]annulene series, those compounds examined show typically anti-aromatic spectra in agreement with the prediction for $4n\pi$ systems ($n = 4$).

1,3,9-Tridehydro-[16]annulene [96,97] 11, (see Table 4) shows a very complex spectrum at room temperature [98-100]. The assignments as shown in Table 4 were aided by frequency swept double irradiation experiments. There exist in this molecule two *trans* double bonds, rotation of each of which could lead to four conformational isomers 13a-d of 11.

---

[1] For the n.m.r. of a nitro derivative (uncertain structure) and of two apparent monomethylsulphonates of monodehydro-[14]annulene, see Ref. [89].

At room temperature, rapid interconversion between these conformers prevents the observation of discrete bands corresponding to inner and outer protons.

13a   *13b*

13c   *13d*

On cooling to −75 °C the interconversion between different conformers is effectively "frozen out" (on the n.m.r. time scale). At this temperature the n.m.r. consists of a superposition of the non-mobile spectra arising from the conformers *13a-d* [98]. From a consideration of the integration, and again with the help of double irradiation experiments, assignments can be made. These assignments lead to the conclusion that the conformers having H(7) in an internal position (*13a* and *13b*) are slightly preferred to those in which H(7) is in an external position (*13c* and *13d*). No conclusion can yet be reached about the favoured conformation of the other *trans* double bond.

There is a further interesting facet of the double irradiation experiments performed on the mixture of conformers at −75 °C. If H(8) in its external environment (τ 2.40) is irradiated, the signal corresponding to the internal H(8) (τ −3.05) disappears. A similar result has been noticed for [18]annulene [101]. This phenomenon arises because the irradiated protons exchange to the alternative environment before relaxation is complete. Thus chemical exchange occurs before equilibrium magnetization is restored, and transfer of saturation results [102]. In this particular case, the disappearance of the signal at τ −3.05 supports the assignment that the signals at τ 2.40 and τ −3.05 are due to the same proton in different sites.

The large difference in the chemical shifts of H(7) (τ 1.10) and H(8) (τ −3.05) in their internal environments has been attributed to long range shielding by the lone triple bond. However, this must be incorrect as the analogous difference for the α- and *cis*-β-protons in vinyl acetylene is predicted to be only 0.15 ppm [49]. Furthermore, planar models indicate that H(7) should be more deshielded by the di-yne system than H(8). It may well be that the observed shift of H(7) represents greater non-planarity in that conformation (*13a* or *b*). This is supported by the rather low value (5 Hz) for $J_{8,9}$ in this conformer.

The room temperature spectrum [66,90,98] of 1,9-didehydro-[16]-annulene [96], **12** (see Table 4), again is of the mobile type, and again the magnetic equivalence of various sets of protons is brought about by the rapid rotation of each of the two *trans* double bonds present in this molecule. At —80 °C the interconversion is slowed down sufficiently for the observation of distinct environments for the two inner protons (τ 0.2) and the ten outer protons (τ 3.92 to 4.55). These chemical shifts are in the regions expected for a molecule possessing a paramagnetic ring current.

The rapid rotation of these bonds should give rise to two non-identical conformers, *14a* and *b* [98]. In fact it has not yet been possible to determine the relative importance of these conformers, as the spectrum of this molecule at —80 °C shows little of the fine structure which made possible the conformational analysis of **11**.

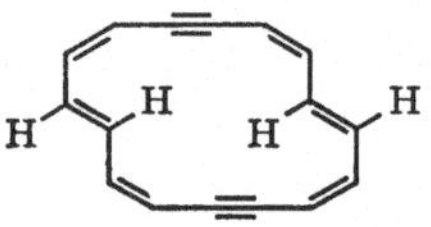

14 a 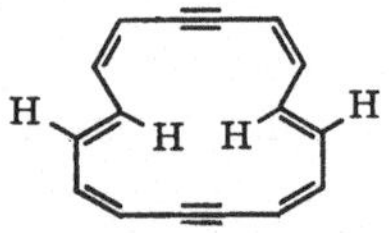   *14 b*

For 1,3-didehydro-[16]annulene [96], **13** (see Table 4), analogous interconversion between non-equivalent conformers occurs. The n.m.r. spectrum [98] was investigated at 0 °C, as the compound is very unstable. Double irradiation again aided in the interpretation of this spectrum. At this temperature it appears that rotation of only two of the three double bonds occurs, leading to four possible conformers, *15a-d* [98]. Thus

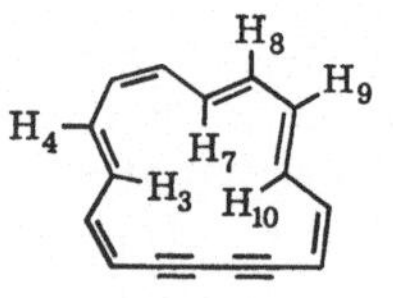

15 a 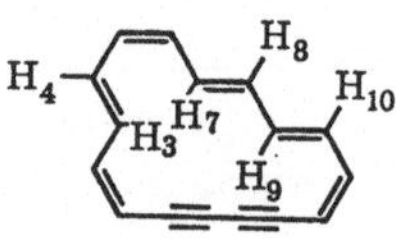   *15 b*

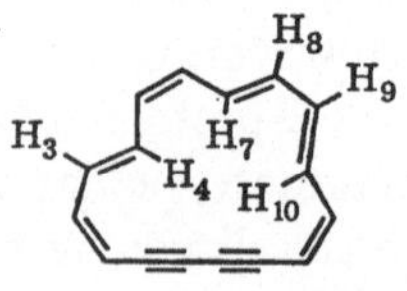

15 c 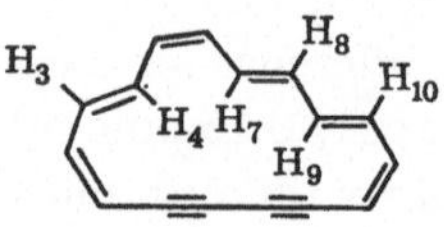   *15 d*

148

H(7) is held in an inner position, and resonantes at τ —3.05 whilst H(8), attached to the same *trans* double bond, resonates at approximately τ 4.8. Using these two chemical shift values as standards, it can be inferred that H(3) and H(10) (τ —0.53 to —0.48) occupy an inner position for about two-thirds of the time, and thus H(4) and H(9) (τ 1.8 to 1.87) occupy an outer position preferentially.

On cooling, changes occur which can be attributed to the slowing down of the interchange between H(3) and H(4) and between H(9) and H(10). However, no further information concerning the relative contributions of the four possible conformers has been obtained.

In considering the interconversion between non-equivalent conformers in **11**, **12** and **13**, only planar forms have been discussed, and the large shifts observed between inner and outer protons for each of these compounds would justify this assumption, at least as a first approximation. In each case the low chemical shift of the inner protons and the upfield shift of the outer protons is consistent with the prediction of anti-aromaticity.

The parent annulene of this series, [16]annulene [96,97,103], **14** (see Table 4) shows a singlet [66] at room temperature at about τ 3.30. As predicted [29], this spectrum changes to the non-mobile type at low temperatures [103]. Thus, at —120 °C the inner protons appear at a very low field (τ —0.32), and the outer protons at τ 4.8.

A quantitative analysis [104] of the n.m.r. spectra of [16]annulene between 0 °C and —150 °C (in carbon disulphide/perdeuteriotetrahydrofuran) has yielded the result that in solution a dynamic equilibrium exists between two configurations of the compound *16a* and *16b* probably via other intermediate conformations.

Considering planar molecules (although X-ray studies [105] have indicated that such a simplification is not necessarily justified) the observation of a singlet at 30 °C can be interpreted as being due to three rapid processes whereby all protons become equivalent [103,104]. Thus, interconversion between *16a* and *16b*, (rapid above —57 °C), rotation of the *trans* double bonds (conformational isomerism) and movement of the π bonds (valence isomerism) combine to allow each proton to experience the same environment (see Fig. 5). At lower temperatures (—67 °C) interconversion between the two configurations *16a* and *16b* is frozen out, and a signal due to each is observed. Broadening of the singlet at τ 3.63 heralds the slowing of the bond rotations in *16a*, whereas a sharp singlet at τ 2.93 attributed to *16b* indicates that the interconversions in this system are more rapid than those in *16a*. At —83 °C the *16b* signal also broadens noticeably, and at very low temperatures (—130 °C) the bond rotations in both *16a* and *16b* are frozen out, and discrete bands due to inner and outer protons can be observed. That

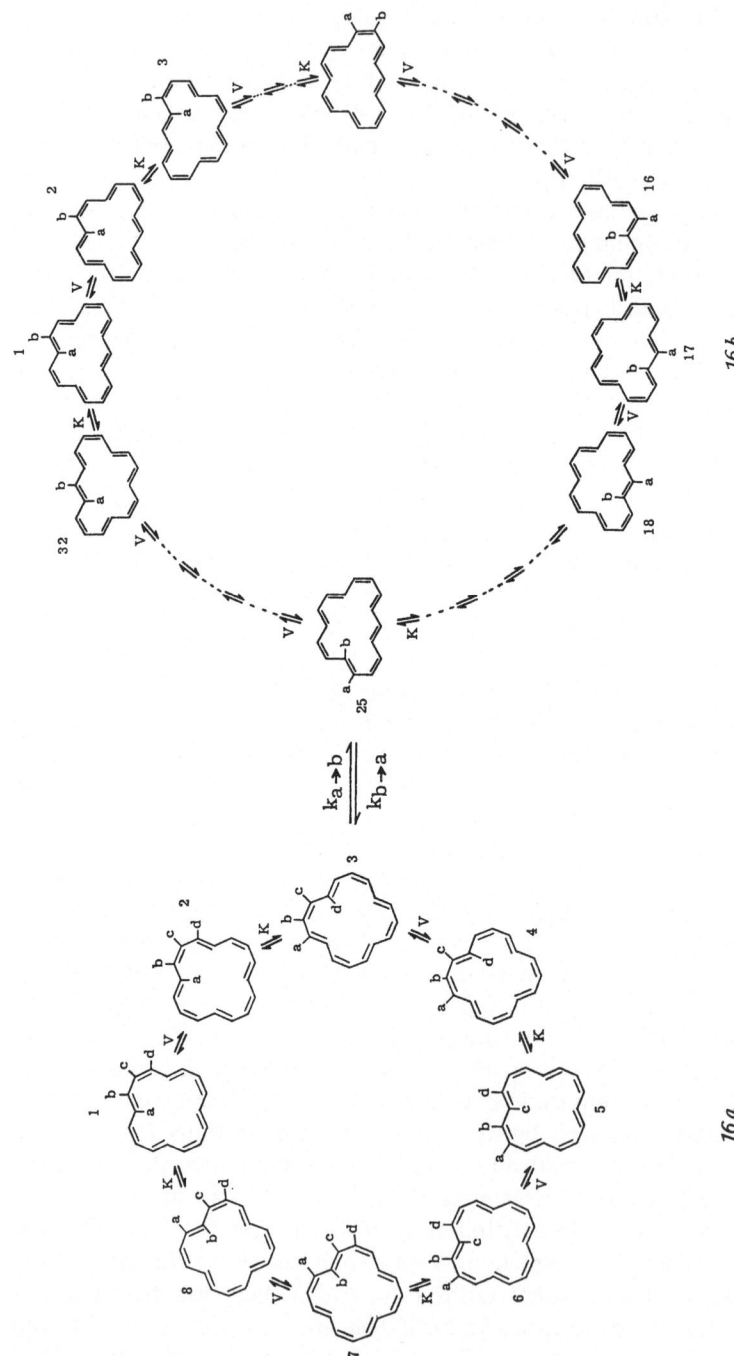

Fig. 5. Mechanisms for interconversion of conformers and isomers of [16]annulene. Processes labelled K and V refer to bond rotation and valence isomerization, respectively. $k_{a \to b} = 17$ sec.$^{-1}$, $k_{b \to a} = 49$ sec.$^{-1}$ $^{104)}$. Reprinted with permission of the copyright owner from Tetrahedron Letters 6259 (1968)

most of the (still broadened) inner and outer proton signals of *16b* are hidden beneath the well-defined bands of *16a* at this temperature, and that *16a* is three times as prevalent as *16b* accounts for the earlier assumption that *16a* was the only configuration present in solution[m) 103)].

The [18]annulenes are a Hückel system, having $(4n + 2)$ π-electrons, where $n = 4$. Furthermore, the ring size is now such that the crowding of internal hydrogen atoms should not be severe enough to cause substantial distortions from planarity. On these two grounds, then, diamagnetic ring currents should cause typically aromatic spectra, and each member of this series so far investigated amply substantiates this prediction.

1,3,7,9,13,15-Hexadehydro-[18]annulene [80)], **15**, (see Table 4) possesses no inner protons and the six equivalent outer protons resonate at τ 2.98. This represents a downfield shift of over 1 ppm from a model compound, *cis*-hex-3-ene-1,5-diyne, *9*, (τ 4.11). This model compound also served to clarify the interpretation of the singlet at τ 5.55 in the case of 1,5,9-tridehydro-[12]annulene, **1**, a molecule for which aromaticity is not predicted. The downfield shift of **15** and the upfield shift of **1** from the acyclic model can then be attributed to a diamagnetic ring current in **15** and a paramagnetic ring current in **1**.

The outer protons of 1,3,7,13-tetradehydro-[18]annulene [107)], **16**, (see Table 4) resonate in a range encompassing the chemical shift of the hexadehydro-[18]annulene, and, as expected, the two inner protons of this compound are shielded, and absorb as two quartets at τ 7.66 and τ 7.80. The structure given in Table 4 is the most likely [107)] of the four structures possible[n)] for a tetradehydro-[18]annulene containing three *cis* and two *trans* double bonds.

1,5,10,14-Tetramethyl-6,8,15,17-tetrahydro-[18]annulene [108)], **17**, is an interesting system for several reasons. It cannot be represented by a conventional formula of conjugated double and triple bonds (c.f. 1,8-didehydro-[14]annulene, **5**) and the equivalence of its four outer protons (τ 0.34 doublet) and its two inner protons (τ 15.24, triplet) shows that it is best represented as the symmetrical structure shown in Table 4. Such a structure can be thought of as representing the resonance hybrid of the two Kekulé structures *17a* and *17b*.

---

m) Studies of the temperature-dependent spectra of two substituted [16]annulenes have also been described [106)].

n) The four structures are possible because the two *cis/trans* butadiene units can each be incorporated into the ring in two ways. This degeneracy is inherent in the preparation by oxidative coupling of 1,5-hexadiyne and subsequent prototropic rearrangement [107)].

17 a   17 b

This non-mobile spectrum is clearly indicative of an aromatic compound, and the various protons retain their separate "identities" over a temperature range of —60 °C to + 64 °C. However, the actual chemical shifts of each of the three types of protons (inner, outer, and methyl) are temperature dependent. Thus the inner protons are shifted *upfield* with decreasing temperature while both the outer H(2) protons and the methyl protons are shifted *downfield*. If the difference in chemical shifts between inner and outer protons is an index of the magnitude of the ring current, it would appear that the ring current increases with decreasing temperature. Such a phenomenon has been observed for [18]annulene [90, 95], 21, and agrees with the finding in that case [109] that the displacement of the atoms from the mean molecular plane decreases as the temperature is decreased. Such confirmatory data is not available in the present case, however.

The possibility of varying the sequence of a given number of *trans* and *cis* double bonds around a ring (within limits set by the positions of the acetylene bonds) has been referred to for 1,3,7,13-tetradehydro-[18]annulene, 16. In the case of 1,7,13-tridehydro-[18]annulene [110] this phenomenon has lead to the isolation of two distinct isomers [107] of this compound, isomer I, 18, and isomer II, 19 (see Table 4).

The n.m.r. of isomer I [91] has been completely assigned, and it indicates that the inner protons resonate at high fields (τ 8.26) and the outer protons are deshielded (τ 1.90 to 2.98). The equivalence of the three inner protons and of the three sets of three outer protons requires the formulation of the head-to-tail arrangement of the double bonds shown in Table 4. The spectrum is unchanged up to 150 °C, a consequence of the fact that a rotation of one the *trans* double bonds would lead to a conformer in which one of the three inner protons suffer considerably increased steric interference.

Examination of the spectrum of isomer II shows non-equivalence of the inner protons, and this compound must therefore be the isomer with a head-to-head linkage of two diene-yne units.

Actually two of the three inner protons are magnetically equivalent and as this is not implied in the structure of 19 it must be due to an accidental equivalence. Assignments to specific inner protons are not feasible as inspection of models reveals little difference in the immediate environments of these three protons. 19, like 18 exhibits a highfield chemical shift for the inner protons, and a deshielding of the outer protons in accord with its expected aromatic nature, as does the 3-nitro derivative [111], 20, of isomer I.

The n.m.r. investigation of [18]annulene [112], 21, (see Table 4) was originally quite difficult as no well defined bands were observed in most solvents [91,92]. The difficulty arises because the room temperature n.m.r. is an intermediate exchange case (somewhere between a mobile and non-mobile spectrum) and as a result the absorptions are so broad as to be almost unrecognizable. Thus the two very broad bands observable in perdeuteriotetrahydrofuran at room temperature sharpen as the sample is cooled, until at −60 °C a well defined quintet at τ 12.99 (six inner protons) and a quartet at τ 0.72 (twelve outer protons) are observable [o].

This non-mobile spectrum shows [18]annulene to be aromatic, a fact which follows from Hückel's theory, together with the finding [113] that this molecule is almost planar, displacement of the carbon atoms from the mean plane being less than 0.1 Å. Further X-ray crystallographic analysis [114] shows that there is no bond alternation in this molecule, (but not all bonds are exactly the same length).

On heating the sample to 110 °C, the bands coalesce to a singlet at τ 4.55, indicating that all protons are experiencing identical average environments, via the same processes which were discussed for [16]-annulene, 14. No mobile spectrum has been observed for any of the dehydro-[18]annulenes as one conformer predominates for each compound.

There appears to be a further n.m.r. temperature dependence for this system similar to that noted for 17, in that the separation between the inner and outer proton bands increases as the temperature is lowered. Thus the separation at −60 °C (12.27 ppm) is larger than that at −20 °C (11.60 ppm) [90,95]. This can be accounted for in terms of an increased ring current at lower temperatures in agreement with the observation [109] that the molecule becomes more planar with decreasing temperature. The magnitude of this effect is larger than that observed for 17, where the separation at −60 °C (15.56 ppm) decreases to 15.30 ppm at −20 °C [108], perhaps indicating that there is a greater tendency towards non-planarity in [18]annulene at room temperature.

---

[o] That the shifts reported [101] for this compound in perdeuteriotoluene are quite different (in particular the inner protons are shifted upfield to τ 14.22) implies strong solvent/solute interactions in this system.

The low temperature spectra [111] of two substituted [18]annulenes, the nitro and the acetyl derivatives (**22** and **23** resp.), are also indicative of aromatic compounds, as they show the characteristic shielding of the six inner protons and the deshielding of the eleven outer protons. On warming these solutions the spectra do not coalesce to a singlet, but rather seem to represent the coalescence of only some of the protons (c.f. the temperature-dependent spectra of some substituted [16]-annulenes [106]). Thus, for **22** about five protons remain permanently in an external position (τ 1.4 to 2.0, area 4.7) whilst the remaining twelve protons are averaging (τ 6.7, broad band, area 12.3). This can be explained in terms of the three possible conformers of a mono-substituted [18]annulene, *18a-c* [111]. If R is a bulky group the conformer in which it

18a   *18b*   *18c*

occupies an internal position will be unfavourable. Thus, the five protons associated with R will not occupy an internal position, and these are the ones which continue to absorb in the low-field region as the temperature is raised. The high temperature spectra of **22** and **23**, then, represent two interchanging conformers, *18b* and *18c* whereas for the unsubstituted [18]annulene, *18* (R = H), all three conformers are possible and a singlet results.

Of the various [20]annulenes prepared [115,116] only one compound has as yet been investigated by n.m.r. This is 1,11-didehydro-[20]-annulene, **24**, whose n.m.r. [66] indicates the presence of a paramagnetic ring current. The room temperature spectrum is of the mobile type arising from the rotation of two pairs of *trans* double bonds, which leads to an averaged "symmetrical" molecule (a similar case was seen for 1,9-didehydro-[16]annulene, **12**). When the bond rotation is "frozen out" at −80 °C, one sees the deshielded inner protons (τ −1.60 to −0.45) and shielded outer protons (τ 4.40 to 4.93) characteristic of an anti-aromatic compound.

The [22]annulene series is predicted to be aromatic [(4n + 2) out-of-plane π-electrons, where n = 5] and to be capable of sustaining a diamagnetic ring current. However no (unbridged) compound in this series has yet been prepared and the predictions remain untested.

1,7,13,19-Tetradehydro-[24]annulene[110,91], 25, exhibits a spectrum[66] typical of an anti-aromatic compound, with its four equivalent inner protons absorbing at τ 1.60, and the outer protons appearing upfield at τ 4.40 to 5.02. Heating to 150 °C does not result in a mobile spectrum, presumably because rotation of any of the four *trans* double bonds would lead to a non-equivalent conformer having more severly hindered inner protons. In this regard 25 is similar to 1,7,13-tridehydro-[18]-annulene, 18. In fact, the rotation of a *trans* double bond adjacent to an acetylene seems to be generally unfavourable, the only example so far found being that of 1,3,9-tridehydro-[16]annulene, 11 [98].

[24]Annulene [112], 26, the two most likely conformations of which are *19a* and *19b* (containing three and four *cis* double bonds, respectively) exhibits a fairly broad singlet at τ 2.75 at room temperature [117]. This arises from the type of averaging processess (bond rotations and bond isomerizations) seen in the case of [16]annulene [103]. Consequently [29] cooling to —80 °C results in the observation of the non-averaged spectrum [117] (see Table 4) where the inner ring protons (τ —2.9 to —1.2) absorb downfield from the outer protons (τ 5.27). This non-mobile spectrum shows [24]annulene to be anti-aromatic, but the experimental integration of the two bands (1:1.86) does not distinguish between the two possible conformations (1:1.67 for *19a* and 1:2.00 for *19b*) and may even indicate the presence of a mixture of the two.

19a   *19b*

A tridehydro-[26]annulene, 27, has been reported [118], but the structure is only one of the many possible, as the positions of the triple bonds and the stereochemistry of the double bonds are as yet undetermined. The n.m.r. consists of a very broad multiplet at τ 2.0 to 4.5, and on cooling to —60 °C no discrete inner or outer proton resonances are observed. It is concluded that this compound, although formally a Hückel $(4n+2)$ system, is not aromatic. As noted previously, the

Hückel criterion for aromaticity becomes insufficient where bond alternation is preferred [2c,7]. The critical size which demarks the onset of bond alternation has been calculated [2c], and is expected to fall between [22]- and [26]-annulene. Hence among the $(4n+2)\,\pi$ annulenes, aromatic character is predicted for values of $n \leqslant 5$, whereas for $n \geqslant 6$ non-aromatic, or polyenic properties are expected. The results reported to date for $(4n+2)\,\pi$ systems seem to be compatible with this prediction, as all of the $n \leqslant 5$ annulenes appear to be aromatic, while a tridehydro-[26]-annulene $(n=6)$ does not[p]. The study of the [22]annulene series $(n=5)$ will serve to demark more clearly the limit at which bond alternation quenches the ring current. Preliminary studies of the n.m.r. of two compounds in the $n=7$ series, viz. a pentadehydro-[30]annulene [110] and a tridehydro-[30]annulene [116], also indicate that their ring size is above the bond-alternation limit, as they both exhibit complex signals in much the same region (τ 2.5 to 4.5) [66] as did 27. The isomeric purity of these last three compounds does not seem to have been unequivocally established, but the observation that all three are non-aromatic is probably sound. In the non-Hückel $4n\pi$ series, in every case where a non-mobile spectrum has been achieved, compounds having $n=3$ to 6 have reflected the type of behaviour expected from compounds which sustain a paramagnetic ring current [29-31].

Coupling constants in this series have proved useful in spectral assignment. Thus, the cis and trans coupling constants across formal double bonds fall in the expected ranges of 10 to 13 and 15 to 16 Hz, respectively. These values are somewhat reduced in cases where resonance is well developed. For example the o-coupling constant in benzene is 7.7 Hz [119] and the cis and trans coupling constants in 1,8-didehydro-[14]annulene, 5, have value of 8.0 and 13.3 Hz, respectively. The coupling constants across formal single bonds will, of course, vary with the dihedral angles, and it is reasonable to suppose that abnormally low values, such as encountered in 1,5-didehydro-[12]annulene, 2, and the conformers 13a and 13b of 1,3,9-tridehydro-[16]annulene, 11, (see p. 147) reflect the degree of non-planarity in such systems.

Table 5 summarizes the chemical shifts of the unsubstituted compounds studied to date (excluding those three having ring sizes above that expected for bond alternation). The range of chemical shifts of inner protons for the $(4\,n+2)\,\pi$ systems in Table 5 is τ 7.66 to 15.48, whereas

---

p) As was pointed out earlier, the difference in energy between the symmetrical and distorted configurations is expected to be rather small in this case, and by introducing three triple bonds into [26]annulene we may have prejudiced 27 towards a bond alternation which would not necessarily be present in the parent compound.

Table 5. *Chemical shifts and Hückel parities for unsubstituted annulenes and dehydroannulenes*

Series	Compound	Chemical shifts (τ)		Hückel parity
		Inner protons	Outer protons	
$(4n+2)\pi$, $n=1$	Benzene		2.734	+
$4n\pi$, $n=3$	1,5,9-Tridehydro-[12]annulene		5.55	—
	1,5-Didehydro-[12]annulene	ca. —7	ca. 4.8 to 5.8	—
$(4n+2)\pi$, $n=3$	1,5,9-Tridehydro-[14]annulene	14.96	0.53 to 1.92	+
	1,8-Didehydro-[14]annulene	15.48	0.36 to 1.46	+
	1,7-Didehydro-[14]annulene	10.7	1.4 to 2.7	+
	Monodehydro-[14]annulene "Stable conformer"	10.7	1.5 to 2.7	+
	Monohydro-[14]annulene. "Unstable conformer"	10.6	1.6 to 2.6	+
	[14]Annulene. "Isomer A"	ca. 10.0	ca. 2.4	+
	[14]Annulene. "Isomer B"	Non-mobile spectrum unachieved		+ (?)
$4n\pi$, $n=4$	1,3,9-Tridehydro-[16]annulene	—3.60 to 1.10	2.40 to 5.6	—
	1,9-Didehydro-[16]annulene	0.2	3.92 to 4.55	—
	1,3-Didehydro-[16]annulene	—3 to —0.5	1.8 to 5.5	—
	[16]Annulene	—0.32	4.8	—
$(4n+2)\pi$, $n=4$	1,3,7,9,13,15-Hexadehydro-[18]annulene		2.98	+
	1,3,7,13-Tetradehydro-[18]annulene	7.66 to 7.80	2.03 to 3.5	+
	1,7,13-Tridehydro-[18]annulene, isomer I	8.26	1.90 to 2.98	+
	1,7,13-Tridehydro-[18]annulene, isomer II	8.17 to 8.28	1.7 to 3.1	+
	[18]Annulene	12.99	0.72	+
$4n\pi$, $n=5$	1,11-Didehydro-[20]annulene	—1.6 to —0.45	4.0 to 5.1	—
$4n\pi$, $n=6$	1,7,13,19-Tetradehydro-[24]annulene	1.60	4.40 to 5.02	—
	[24]Annulene	—2.9 to —1.2	5.27	—

their outer protons appear in the range τ 0.36 to 3.5. Conversely the 4 $n\pi$ systems show a relative deshielding of the inner protons (τ *ca.* —7 to 1.60), whilst their outer protons resonate at higher fields (τ 2.4 to 5.8).

In summary then, it appears that all of the monocyclic, unbridged annulenes and dehydroannulenes so far investigated do indeed show the predicted behaviour with regard to "Hückel aromaticity" (within the ring size limit). Throughout the study of these compounds the question of steric interference of the inner protons has been an important one (as deviations from planarity are expected to quench the ring current to some degree). This consideration has not been critical in the large rings, but becomes more severe with the smaller rings, and is a contributing cause in the difficulty experienced in isolating [12]annulene and [10]-annulene (see however Addendum, p. 208).

F. N.M.R. Spectra of the Quasi-Annulenes

In the preceding section we have included those annulenes of general structural formula C_nH_n (n even) and their dehydro analogues. There are however, other compounds which are in the spirit of annulenes and to a greater or lesser extent fit within the purview of this review. These "quasi-annulenes" may conveniently be considered to fall into three subgroups.

Perhaps the most important class is comprised of those compounds which have come to be known as bridged annulenes. Many of these compounds bear formal analogy to cata-condensed benzenoid hydrocarbons with the exception that any bond common to two or more ring systems is replaced by a non-conjugating bridge group (hence retaining the concept of a single conjugated cycle of carbon atoms). In addition to the property of π-insulation expected of a bridge group, it is also expected that the hybridization and bond lengths will be such that the perimeter π-system will, as far as possible, be able to assume its preferred, presumably planar configuration. For many of the compounds considered here it is obvious that the bridging group does not interact electronically with the perimeter π-systems, and this observation may often be corroborated by spectral evidence. In less well defined cases, the decision to include a compound usually indicated that bridge-perimeter interactions are expected to be minimal. However this assertion is not well founded in the case of some macrocycles discussed below (which also lie outside our stated aims in that they contain some hetero-atoms in the path of conjugation), thus to some extent the decision to include a particular compound reflects the authors' own emphasis. It should be stressed that the bridging group is not necessary to hold these molecules together, rather it is inserted to prevent the molecules from being drastically deformed by the abuttal of internal hydrogen atoms; thus the bridge absorbs the free valence created by the removal of the offending protons.

In the second class we discuss the aromatic ions, which are of general formula $C_{(4n+1)}H^-_{(4n+1)}$ or $C_{(4n+3)}H^+_{(4n+3)}$ ("Type 1") and $C_{4n}H^{2+}_{4n}$ or $C_{4n}H^{2-}_{4n}$ ("Type 2"). Although the HMO method does not strictly apply to charged species it has again qualitatively indicated the correct direction of investigation. All the ions possess a closed shell configuration

Type	Type 1								Type 2			
Ion	Cyclopropenyl C_3H_3		Cyclopentadienyl C_5H_5		Cycloheptatrienyl C_7H_7		Cyclononatetraenyl C_9H_9		Cyclobutadienyl C_4H_4		Cyclooctatetraenyl C_8H_8	
Electronic Charge	Cation	Anion[b]	Cation[b]	Anion	Cation	Anion[b]	Cation[b]	Anion	Di-cation[b]	Di-anion[b]	Di-cation[b]	Di-anion
HMO π-electron Energy	$4\,\beta$	$2\,\beta$	$5.24\,\beta$	$6.47\,\beta$	$8.99\,\beta$	$8.10\,\beta$	$10.82\,\beta$	$11.52\,\beta$	$4\,\beta$	$4\,\beta$	$9.66\,\beta$	$9.66\,\beta$
Resonance Energy I[c]	$2\,\beta$	0	$1.24\,\beta$	$2.47\,\beta$	$2.99\,\beta$	$2.10\,\beta$	$2.82\,\beta$	$3.52\,\beta$	$2\,\beta$	$2\,\beta$	$3.66\,\beta$	$3.66\,\beta$
Resonance Energy II[d]	$1.17\,\beta$	$-0.83\,\beta$	$-0.22\,\beta$	$1.01\,\beta$	$0.93\,\beta$	$0.04\,\beta$	$0.19\,\beta$	$0.89\,\beta$	$0.76\,\beta$	$0.76\,\beta$	$0.84\,\beta$	$0.84\,\beta$
SCF–MO Resonance Energy II (e.V.)[e]	-1.84	1.56	0.73	-1.31	-0.93	0.25						

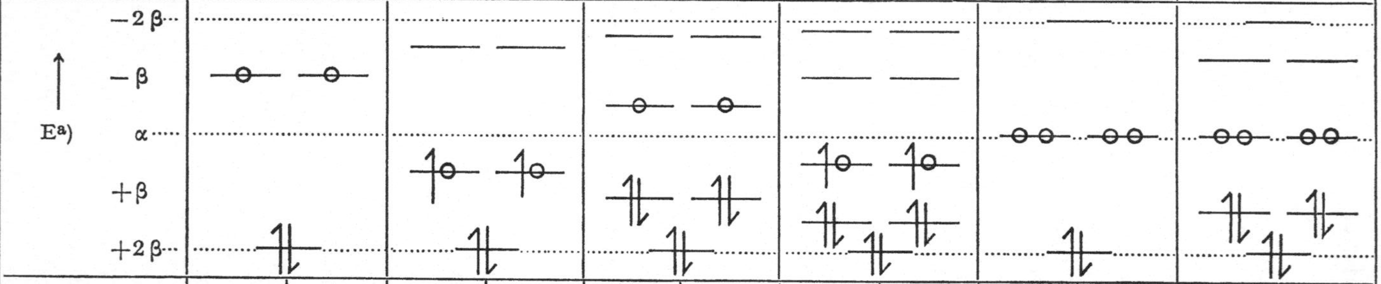

a) β, the carbon-carbon resonance integral, is negative.
b) Hypothetical cases.
c) Defined against appropriate number of non-interacting double bonds [4].
d) Defined against corresponding ionic linear polyene [2,3].
e) Dewar, M. J. S., Venier, C. G., in: Dewar, M. J. S.: The Molecular Orbital Theory of Organic Chemistry, p. 187. New York: McGraw-Hill 1969.

Fig. 6. Hückel π-electron molecular orbitals for some charged annulenes

with the π-electrons occupying low lying molecular orbitals. The π-energy levels for the singly charged species are somewhat analogous to those of the neutral aromatic annulenes (no anti-aromatic ions with unit charge have yet been studied by n.m.r.), and their shielding constants are expected to be subject to the same type of effects (in addition to the perturbations associated with the full charge they bear), hence their inclusion is mandatory.

Of recent interest is the class known as homoaromatic compounds which contain one or more sp^3 hybridised carbon atoms in an otherwise conjugated cycle. These compounds apparently prefer to take up a conformation such that the conjugating segments of the molecule may overlap to some extent, thus "completing" the π-cycle, presumably in order to gain resonance stabilisation and efficient dispersal of charge (all the well authenticated examples of this phenomena bear a residual positive or negative charge). Naturally this homo-conjugation between segments will be characterised by a considerably smaller overlap integral than is normally appropriate for π bonds in carbocyclic systems. For efficient interaction of the conjugated segments, coplanarity and proximity are at a premium and the unhappy methylenic carbon is forced to lie almost vertically above the "periphery" of any resulting cycle. Thus a convenient probe exists in these molecules and the resulting n.m.r. spectra are quite characteristic.

1. Bridged Annulenes

10π-Systems

It has already been noted that planarity in di-*trans*-[10]annulene (*1*, p.112) is precluded by the overlap of the van der Waals radii of the two internal hydrogen atoms. Replacement of these two hydrogen atoms, by a bridging methylene group, however, removes this steric inhibition, and the C_{10} skeleton might be expected to approach planarity. This interesting concept lead Vogel and his coworkers to synthesize a series of bridged 10 and 14 π-electron systems, which in most cases have n.m.r. spectra typical of that expected for $(4n+2)\pi$ annulene systems.

When the parent compound of the bridged [10]annulene series, 1,6-methano-[10]annulene [11,120], *28* (see Table 6), was first prepared, it was necessary to distinguish between three possible structures: *a*) the "double norcaradiene" structure *20*, *b*) a rapidly fluctuating system (c.f. cyclooctatetraene), possibly proceeding through low equilibrium concentrations of the double norcaradiene, *21*, or *c*) a delocalized 10 π-electron system *22*.

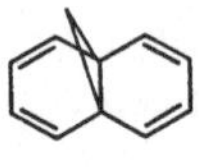

20

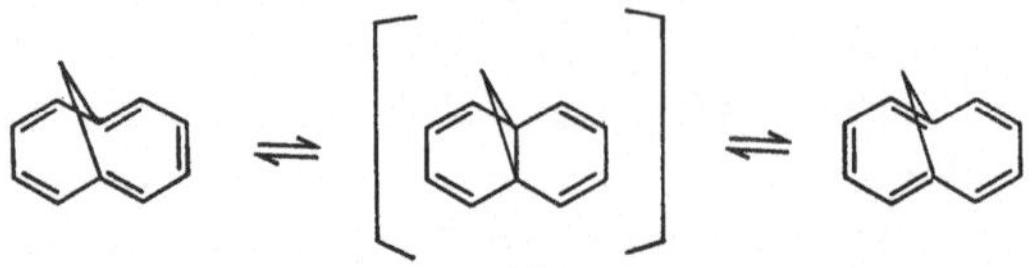

21

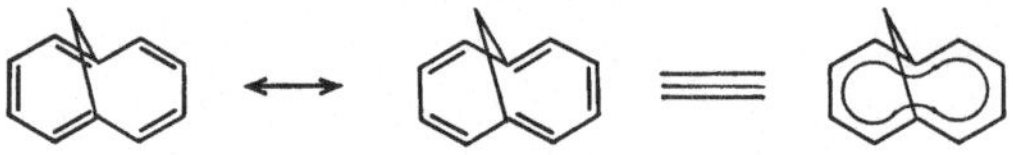

22

The n.m.r. of this compound [11,17,121] establishes the delocalized nature of the 10 π-electron system *22*, and it can be rightly considered as a [10]annulene. Thus the eight peripheral hydrogens absorb not in a typically olefinic region (as would be expected for *20*), but are deshielded (AA′ BB′ system centered at τ 2.9) while the two protons of the bridging methylene are shielded (τ 10.5). The double norcaradiene structure is further mitigated against by a consideration of the ^{13}C-H coupling constant for the bridging methylene group. This is 142 Hz, whereas a typical value for a CH_2 group incorporated into a cyclopropane ring is 160 Hz. That the spectrum is unchanged from room temperature to −140 °C [17] makes the rapid interconversion *21* very unlikely, and the C_{2v} symmetry of the molecule, as reflected by its n.m.r. spectrum, rules out the unlikely possibility of a structure with fixed double bond positions. There is some coupling between the bridge protons and the four α-protons as shown by double irradiation experiments [121]. Irradiation at the frequency of resonance of the α-protons results, then, in a sharpening of the methylene proton signals from a width at half height of 2.0 Hz to 0.7 Hz.

Consideration of 9,10-ethano-9,10-dihydronapththalene, *23* [122,123], a reasonable model for *20*, further supports the delocalized nature of 1,6-methano-[10]annulene. The AA′BB′ system of olefinic protons in *23*

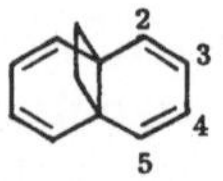

23

τ 4.50 (8 protons) $J_{2,3} = 9.71$ Hz
τ 7.45 (4 protons) $J_{3,4} = 5.48$ Hz
                      $J_{2,4} = 0.72$ Hz
                      $J_{2,5} = 1.09$ Hz

Table 6*. *Bridged* 10π *systems*

Compound	Chemical shifts (τ)	Coupling Constants (Hz)	Solvent, temperature	Ref.
H_{11a} H_{11b} (structure) 2 3 4 5 **28** 1,6-Methano-[10]annulene	2.727 ± 0.001 (H_2, H_5) 3.051 ± 0.001 (H_3, H_4) 10.519 ± 0.002 (H_{11})	$J_{2,5} = 1.46 \pm 0.04$ $J_{3,4} = 9.19 \pm 0.04$ $J_{2,3} = 8.97 \pm 0.03$ $J_{3,5} = 0.02 \pm 0.03$	CCl_4 (1 M solution) R.T.	[121]
	2.896 ± 0.001 (H_2, H_5) 3.182 ± 0.001 (H_3, H_4) 10.486 ± 0.002 (H_{11})	$J_{2,5} = 1.47 \pm 0.04$ $J_{3,4} = 8.97 \pm 0.05$ $J_{2,3} = 8.95 \pm 0.04$ $J_{3,5} = 0.01 \pm 0.05$	Neat *ca.* 50 °C	
		$J_{13C-H_{11}} = 142$[120] $J_{H_{11a}, H_{11b}} = 6.91 \pm 0.26$[a]		
CH_2 (structure) 2 3 4 5 **29** 11-Methylene-1,6-methano-[10]annulene	2.585 (H_2, H_5) 2.984 (H_3, H_4) 6.808 (CH_2)	$J_{2,3} = 8.85$ $J_{3,4} = 9.43$ $J_{2,4} = 0.01$ $J_{2,5} = 1.32$	CCl_4	[135]
O (structure) 2 3 4 5 **30** 1,6-Oxido-[10]annulene	2.541 ± 0.001 (H_2, H_5) 2.744 ± 0.001 (H_3, H_4)	$J_{2,5} = 1.13 \pm 0.05$ $J_{3,4} = 9.28 \pm 0.05$ $J_{2,3} = 8.77 \pm 0.04$ $J_{2,4} = 0.28 \pm 0.03$	CCl_4	[121]

* See footnote to Table 4, p. 134.
a) Coupling constant inferred from compound monodeuterated in the bridge position [121].

Table 6 (continued)

Compound	Chemical shifts (τ)		Coupling constants (Hz)	Solvent, temperature	Ref.
31 1,6-Imino-[10]annulene	2.592 2.889 ~ 11.2	(H$_2$, H$_5$) (H$_3$, H$_4$) (NH)	$J_{2,3} = 8.82$ $J_{3,4} = 9.31$ $J_{2,4} = 0.06$ $J_{2,5} = 1.50$	CCl$_4$	135)
32 N-Methyl-1,6-imino-[10]annulene	2.734 2.997 9.415	(H$_2$, H$_5$) (H$_3$, H$_4$) (CH$_3$)	$J_{2,3} = 8.94$ $J_{3,4} = 9.08$ $J_{2,4} = 0.11$ $J_{2,5} = 1.47$	CCl$_4$	135)
33 Cycl[3,2,2]azine	2.80 2.49 2.14 2.41	(H$_1$, H$_4$) (H$_2$, H$_3$) (H$_5$, H$_7$) (H$_6$)	$J_{1,2} = 4.2 \pm 0.2$ $J_{5,6} = 8.0$	CCl$_4$	148, 149)

resonate at τ 4.5, and the bridging CH_2 protons resonate as low as τ 7.45. Thus the unusually low-field olefinic absorption and the high-field methylene absorption for 1,6-methano-[10]annulene arises from the presence of a diamagnetic ring current in this system. The fact that this molecule can be considered aromatic but that a stable unbridged [10]annulene is unknown demonstrates anew the profound effects which steric requirements exert on the electronic nature of the ring.

Various substituted 1,6-methano-[10]annulenes have been prepared and characterized, and in particular the n.m.r. of such compounds reflects their delocalized nature [17,120,124-127] (a π-complex formed from reaction of 1,6-methano-[10]annulene with hexacarbonylchromium has also been characterized [128]). The X-ray crystallographic analysis [129] of the 2-carboxylic acid derivative 24 shows the bridge angle to be 99.6°, and thus that the C_{10} perimeter is even flatter than expected from molecular models. Furthermore, the ten perimeter bonds have lengths which vary over the narrow range 1.38 to 1.42 Å, the values being typical of benzenoid aromatic bonds. This lack of appreciable

COOH   COOCH₃

24   *25*

bond alternation is also demonstrated by the similarity of the vicinal coupling constants $J_{1,2}$ and $J_{2,3}$ (see Table 6) [121].

The methyl ester of this compound, 25, has served to clarify the question as to whether the bridge can flip through the plane of the carbon perimeter in this series of compounds, as an optically active sample of 25 would be racemized by such a flipping process. That optical activity is retained even up to 250 °C shows the absence of bridge flipping in 25, and, by inference, in the unsubstituted compound [130].

1,6-Methano-[10]annulene undergoes acid catalyzed deuteration to yield the α-d_4 compound 26, in which the AA'BB' system is now a singlet due to the remaining four β-protons. The bridge proton singlet has sharpened, indicating (as did double irradiation experiments [121])

D   D   D   H

                        methylene proton
                        τ 10.5 (*t.*)
D   D $J_{H,D} = 1.06 \pm 0.04$ Hz.

26   *27*

that a small α-proton to bridge proton coupling occurs in the undeuterated parent hydrocarbon [17]. The d_0, d_1, and d_3 intermediate cases have also been prepared and studied [131].

The derivative, 27, mono-deuterated in the bridge position has been synthesized, and the observed geminal $J_{H,D}$ for this system of 1.06 Hz corresponds to a value for the geminal methylene coupling in the undeuterated compound of 6.91 Hz. This value for a geminal coupling is similar to that observed for bridges in norbornene and benzonorbornadiene and is substantially larger than the values for cyclopropane methylene groups [121].

The bridge protons of 1,6-methano-[10]annulene are formally allylic, but this position is unreactive. Such unreactivity is due to the fact that any $p\pi$ orbital which might be formed at the bridge would be orthogonal to the p_z orbitals of the perimeter π-system, and consequently would not be resonance stabilised. The study of bridge substituted compounds could only be undertaken by introducing substituents at an earlier stage of the synthesis [17], and, starting with 11-bromo-1,6-methano-[10]annulene, an extensive series of such derivatives have been prepared and studied [132–134].

Of particular interest is 11-methylene-1,6-methano-[10]annulene [17, 132,135], 29 (see Table 6). The olefinic bridge protons are shifted to unusually high field (τ 6.8), reflecting substantial shielding by the 10 π-system. However, the parameters of the AA'BB' perimeter system remain similar to those of 1,6-methano-[10]annulene, and the ultraviolet spectra of 28 and 29 are almost identical, indicating that the 10 π-electron system of 29 is not perturbed by interaction with the π-electron system of the bridge. The bridge methyl protons of 11-methyl-1,6-methano-[10]annulene [17], 28, also provide a probe for a shielding effect in the region above the 10 π-electron system, and once again an upfield shift (τ 10.3) is observed [136].

H CH₃

τ 10.3 (4 protons)
τ 3.1 (8 protons)

28

A final sub-group within the bridged 10 π-electron systems includes those compounds bridged by heteroatoms. The synthesis and n.m.r. investigation of 1,6-oxido-[10]annulene [137–139,142] and many of its derivatives [17,125,138,140] have accordingly been executed. The n.m.r. spectrum [17,121] of 1,6-oxido-[10]annulene itself (30, see Table 6) consists of an AA'BB' system centered at τ 2.65 (in CCl₄; also quoted at τ 2.52

in CDCl$_3$ [138]), being shifted downfield by approximately 0.2 ppm from the 1,6-methano-[10]annulene. Definite assignments to the α- and β-protons in the AA′BB′ system cannot be made from a consideration of the coupling constants $J_{A,A'}$ and $J_{B,B'}$, since the calculated spectrum is insensitive to interchange of their values (1·1 and 9·3 Hz) [121]. However, the spectrum of 1,6-oxido-[10]annulene-2,5,7,10-d_4 establishes that the less shielded protons are at the α-positions. This is in agreement with a prediction based on Pople's point dipole approximation [52] when applied in the same manner as for naphthalene [131]. The u.v. spectrum of the oxido analogue **30** is almost identical to that of the parent hydrocarbon **28**, implying no interaction between the lone pairs of electrons on oxygen and the 10 π-electron system [139]. The X-ray study of this compound [141] shows its perimeter to be more planar than that of 1,6-methano-[10]-annulene-2-carboxylic acid, *24*, with the eight proton-bearing carbons having a root mean square deviation from the mean plane of only 0.025 Å, and the C(1) and C(6) atoms being 0.35 Å above this plane. The carbon-to-carbon bond lengths all lie within a typically aromatic range close to 1.39 Å. This lack of bond alternation is further manifest in the almost identical values of the 1,2 and 2,3 vicinal coupling constants (8.8 and 9.3 Hz, resp.) [121].

The nitrogen bridged analogue, 1,6-imino-[10]annulene, **31** (see Table 6), also possesses an n.m.r. spectrum which reflects the delocalized nature of the 10 π-electron system [143,135]. The eight peripheral protons appear as an AA′BB′ system centered at τ 2.8, the C_{2v} symmetry of the C$_{10}$ perimeter being the result of a rapid inversion of the nitrogen atom [135]. Further indication of a diamagnetic current in **31** is the strong shielding of the imino proton which resonates at τ 11.2, and again the u.v. spectrum adds proof that the methylene, oxido and imino bridged [10]annulene compounds have very similar delocalized 10 π-systems [143].

Various N-substituted derivatives of **31** have been synthesized, and their n.m.r. spectra investigated [143,139]. In particular the n.m.r. of the N-methyl compound [135], **32** (see Table 6), shows the shielding of the methyl protons (τ 9.4) expected for a group situated above a delocalized 10 π-electron system. That the methyl resonance is a sharp singlet, and that the vinylic protons still appear as a symmetrical AA′BB′ system necessitates a process by which the methyl group "sees" both sides of the molecule equivalently. This process could be a rapid inversion at the nitrogen [135,143] as is seen for **31**. Alternatively, it has been suggested [144] that the nitrogen is sp^2 hybridized and the methyl group is thereby held in a position symmetrical with respect to the two half-rings. Low temperature n.m.r. has not distinguished between these two possibilities, and experiments with asymmetrically substituted derivatives have, to date, proved inconclusive [135].

The spectral parameters of compounds **28** to **32** are all very similar, and dramatic changes in the bridging group throughout the series do not significantly affect the perimeter coupling constants. The molecular geometry thus does not change much throughout the series [129,135,141]. The bridging unit enables a fair degree of planarity to be attained and thus allows delocalization of the 10 π-electrons to take place. Such planar, $(4n+2)\pi$ systems (where $n=2$) are expected, by our earlier definition, to be aromatic in nature, and the n.m.r. spectra of these compounds have amply reflected the presence of a diamagnetic ring current in each case. Furthermore, the perimeter *ortho* coupling constants do not vary largely within each compound, indicating a strong tendency toward bond equalization throughout the series [65,121,129].

Another C_{10} perimeter which becomes feasible with suitable bridging is the structure **29** possesssing three *trans* double bonds. In this case the unfavourable steric interaction of the three internal hydrogen atoms

29

would be circumvented by replacing them with a trigonally hybridized bridging group, preferably one whose inclusion in the molecule will not perturb the peripheral π-system. Cycl[3,2,2]azine [145,146], **33** (see Table 6), where the system is bridged by a nitrogen atom, appears to satisfy this condition. Calculations [147] have shown that the extent of delocalization into the perimeter π-system of the lone pair of electrons on nitrogen is small. That the perimeter system of **33** can be considered as an annulene is seen by its n.m.r. spectrum [148,149].

Thus the protons resonate in the range τ 2.1 to 2.8, which is typically aromatic. It should be realized that any delocalization of the nitrogen lone pair electrons will cause an upfield shift in the proton absorptions, and accordingly the low field resonances observed must reflect a substantial diamagnetic ring current.

12π-Systems

The next highest homolog of **33** is cycl[3,3,3]azine. This compound has been synthesized [150] and shown to be a good model for the unknown [12]annulene, *30* (see however Addendum, p. 208), the three internal

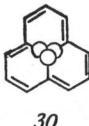

30

hydrogen atoms of which are replaced by a bridging trigonal nitrogen atom (**34**, see Table 7). Thus the protons absorb as an A_2B system at τ 6.35 and τ 7.93 [150]. These remarkably high field shifts both reflect and magnify the trend noted for the unbridged dehydro-[12]annulenes. It should be noted that these shifts cannot be accounted for solely in terms of a delocalization of the nitrogen lonepair electrons into the periphery of the ring, since they correspond to an electron density greater than that which would result if the two electrons were totally distributed amongst these positions.

Table 7*. *A bridged* 12π *system*

Compound	Chemical shifts (τ)	Coupling constants (Hz)	Solvent	Ref.
34 Cycl[3,3,3]azine	6.35 (t., H_2, H_5, H_8) 7.93 (d., H_1, H_3, H_4, H_6, H_7, H_9)	$J_{1,2} = ca.\ 8$	bis(tri-methyl-silylether)	150)

* See footnote to Table 4, p. 134.

So it is that the two cyclazines, **33** and **34**, formally similar in structure, have n.m.r. spectra dominated by the ring currents which are surely present in their peripheral π-systems; a diamagnetic current in the case of the [10]annulene-like cycl[3,2,2]azine and a paramagnetic current in the [12]annulene-like cycl[3,3,3]azine.

14π-Systems

In the bridged 14 π-electron systems, as in the 10 π-bridged compounds, there is the possibility of synthesizing molecules having different perimeter geometries. The pyrene-type perimeter *31* requires the involvement of four perimeter carbon atoms in bridging to eliminate the steric crowding which four inner protons at these positions would introduce. Four

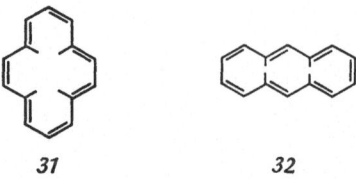

31   *32*

internal protons must also be replaced in a bridged [14]annulene having the anthracene-like perimeter *32*; in this case two independent bridges could render the system almost planar. A perimeter of type *31* is that assigned to [14]annulene itself, which, it was seen, exists as two distinct conformers (**9** and **10**, see Table 4). These conformers arise because the four inner protons cannot be accommodated in one plane (see however Addendum, p. 209). The bridged [14]annulenes synthesized by Boekelheide and coworkers possess this perimeter bridged by an internal 2,3-butano, 3,4-hexano, or 4,5-octano unit, and provide a clue to the nature of a (hypothetical) planar [14]annulene (see also Addendum, p. 209).

Trans-15,16-dimethyldihydropyrene [151,152], **35** (see Table 8), has perimeter protons which are well deshielded, appearing in a range of τ 1.33 to 1.86. Assignment of absorptions to particular protons is simplified by the study of *33*, the 2,7-dideuterated derivative of **35** [151,152]. A multiplet centered at τ 1.86 disappears on deuteration, and can thus be

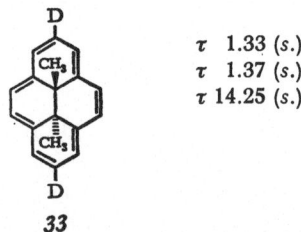

τ 1.33 (*s.*)
τ 1.37 (*s.*)
τ 14.25 (*s.*)

33

assigned to protons 2 and 7. The singlet due to H(4), H(5), H(9) and H(10) remains unchanged and two peaks at τ 1.33 and 1.43 in the parent compound become a singlet at τ 1.37 on deuteration, indicating that in **35** the 1,3,6 and 8 protons are the A_2 part of an A_2B system, being split by protons 2 and 7.

Table 8*. *Bridged* 14π *systems* [c])

Compound	Chemical shifts (τ)	Coupling constants (Hz)	Solvent	Ref.
35 *trans*-15,16-Dimethyl-dihydropyrene	1.33 (*s.*, H$_4$, H$_5$, H$_9$, H$_{10}$) 1.37 (*d.*, H$_1$, H$_3$, H$_6$, H$_8$) 1.86 (*m.*, H$_2$, H$_7$) 14.25 (*s.*, CH$_3$)	$J_{1,2}=7.5$ $(=J_{2,3}=J_{6,7}=J_{7,8})$	CDCl$_3$	[152]
36 *trans*-15,16-Diethyl-dihydropyrene	1.36 (*s.*, H$_4$, H$_5$, H$_9$, H$_{10}$) 1.33 (*d.*, H$_1$, H$_3$, H$_6$, H$_8$) 2.05 (*t.*, H$_2$, H$_7$) 11.86 (*t.*, CH$_3$) 13.96 (*q.*, CH$_2$)	$J_{1,2}=7$ $(=J_{2,3}=J_{6,7}=J_{7,8})$ J_{vic} (ethyl group) $=7.5$	CDCl$_3$	[156]
37 *trans*-15,16-Di-*n*-propyl-dihydropyrene	1.33 (*s.*, H$_4$, H$_5$, H$_9$, H$_{10}$) 1.33 (*d.*, H$_1$, H$_3$, H$_6$, H$_8$) 2.05 (*t.*, H$_2$, H$_7$) 10.65 (*t.*, γ—CH$_3$) 11.82 (*m.*, β—CH$_2$) 14.03 (*m.*, α—CH$_2$)	$J_{1,2}=7$ $(=J_{2,3}=J_{6,7}=J_{7,8})$	CDCl$_3$	[157, 230]

* See footnote to Table 4, p. 134.

Table 8 (continued)

Compound	Chemical shifts (τ)	Coupling constants (Hz)	Solvent	Ref.
38 *syn*-1,6; 8,13- Dioxido- [14]annulene	2.06 (*s.*, H_7, H_{14}) 2.25 (H_2, H_5, H_9, H_{12}) 2.40 (H_3, H_4, H_{10}, H_{11})	$J_{2,3} = 9.0$ $J_{3,4} = 9.2$ $J_{2,4} = 0.3$ $J_{2,5} = 1.1$	$CDCl_3$	160)
39 *syn*-1,6-Methano- 8,13-oxido- [14]annulene	11.39 (H_{16}) 9.08 (H_{15}) 2.346 (H_2, H_5) 2.664 (H_3, H_4) 2.358 (H_9, H_{12}) 2.571 (H_{10}, H_{11}) 2.254 (H_7, H_{14}) 2.375 (H_2, H_5) 2.700 (H_3, H_4) 2.402 (H_9, H_{12}) 2.631 (H_{10}, H_{11}) 2.290 (H_7, H_{14})	$J_{15,16} = 10.58$ $J_{15,2} = J_{15,5} = 1.15$ $J_{2,3} \;\; = 8.82$ $J_{3,4} \;\; = 9.28$ $J_{9,10} = 9.44$ $J_{10,11} = 9.05$ $J_{2,3} \;\; = 8.72$ $J_{3,4} \;\; = 9.55$ $J_{9,10} = 9.42$ $J_{10,11} = 9.30$	$CDCl_3$ CCl_4	159, 162, 248)
40 1,6; 8,13-Propano- [14]annulene	2.264[a] (H_2, H_5, H_9, H_{12}) 2.446[a] (H_3, H_4, H_{10}, H_{11}) 2.115[a] (H_7, H_{14}) 10.611[a] (2H, H_{16}) 11.157[a] (H_{15}, H_{17})	$J_{2,5} \;\; = 1.37$[b] $J_{3,4} \;\; = 9.50$[b] $J_{3,5} \;\; = 0.41$[b] $J_{4,5} \;\; = 9.15$[b] $J_{15,16} = 2.84$[b]	$CDCl_3$	165)

[a]) Estimated error less than 0.002 ppm.
[b]) Estimated error less than 0.1 Hz.
[c]) See also Addendum, Table 17, p. 211.

Without doubt, however, the most striking aspect of the n.m.r. spectra of these two compounds is the pronounced shielding of the methyl protons of the bridge, which absorb as a sharp singlet at τ 14.25. These methyl groups, which extend into the cavity of the π-electron cloud, are a sensitive probe to the magnetic field arising from the diamagnetic current in the periphery. The planarity necessary for significant delocalization is demonstrated by the X-ray study [153] of 2,7-diacetoxy-*trans*-15,16-dimethyldihydropyrene, *34* [154]. The perimeter atoms are approximately coplanar, the maximum deviation of these atoms from a mean plane

OCOCH$_3$

CH$_3$

CH$_3$

OCOCH$_3$

34

being not greater than 0.027 Å. The bonds of the perimeter have values which vary between approximately 1.386 and 1.401 Å, which is in the range of typical benzenoid aromatic bonds. The n.m.r. spectrum of this compound [152,154], as well as those of other substituted derivatives of *35* [155], reflect deshielding of the peripheral protons, and the large characteristic upfield shift of the bridge methyl groups.

Two higher homologues, *trans*-15,16-diethyldihydropyrene [156], *36*, and *trans*-15,16-di-*n*-propyldihydropyrene [157,230], *37* (see Table 8), similarly possess ring currents as evidenced by the low-field absorption of their perimeter protons (τ 1.33 to 2.05) and the strong shielding of the ethyl and *n*-propyl protons. In *36* and its various derivatives [156] the methylene protons of the ethyl group are shifted to higher fields than the methyl protons (τ 13.96 and τ 11.36, respectively for *36*).

Similarly, in *37* a decrease in shielding occurs on progression from the α-CH$_2$ group (τ 14.03) to the β-CH$_2$ (τ 11.82) and the γ-CH$_3$ (τ 10.65). This effect reflects the changing profile of magnetic field with position. Once again X-ray studies on the diethyl compound [158] indicate an almost planar perimeter, where the maximum deviation from a mean plane (0.081 Å) is greater than that for the dimethyl derivative, as expected for a compound having a bulkier internal substituent (the peripheral bond lengths vary between 1.393 and 1.401 Å).

The other series of bridged 14 π-electron compounds comprises those systems with anthracene perimeter (e.g. *32*, a system which, if allowed to attain planarity, is also predicted to have aromatic character). These compounds can be considered as the next highest members in the series of bridged annulenes based on linearly fused hydrocarbons, extending the study of the naphthalene-perimeter bridged [10]annulenes. There is in the system *35* a possibility of geometrical isomerism, in that the

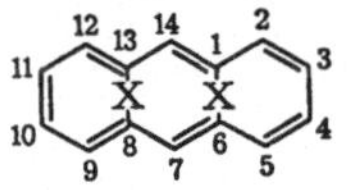

35

two bridges can be *syn* or *anti* with respect to each other. Substantial planarity of the ring perimeter is possible only with the *syn* form.

Equally important in determining the possibility of aromaticity in these systems is a consideration of the orbital overlap between the $2p\pi$ orbitals of C(6), C(7) and C(8) (and between C(1), C(14) and C(13)). In the *syn* isomer substantial overlap is possible, as the orbitals can become almost parallel (*36*). In contrast, the $2p\pi$ orbitals of C(7) and C(8) in the *anti* form are twisted out of alignment, as are those of C (6) and C (7), (*37*). This decrease in overlap would inhibit resonance stabilization

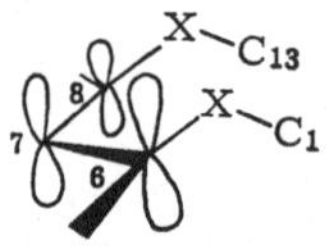

36

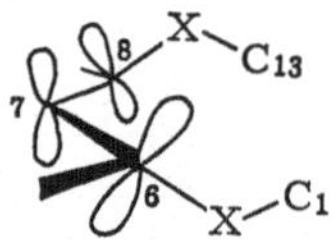

37

in the *anti* form. However, in the *syn* form there is a possibility of serious steric interaction between the two bridges. Thus, it has not yet been possible to synthesize *syn*-1,6;8,13-dimethano-[14]annulene, *36*, due no doubt to the interference of the protons on the bridging methano groups.

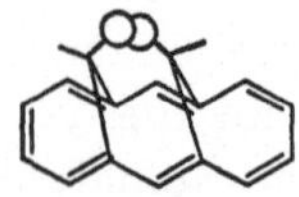

38

The n.m.r. of the *anti* isomer of 1,6; 8,13-dimethano-[14]annulene [127,159,244b,247], *39*, does reflect the olefinic character expected from a consideration of its geometry. The perimeter protons absorb in a region typical of a cyclic polyolefin (τ 3.6 to 3.8) whilst the bridging methylene

39

τ 3.67 (*s.*, 2 protons)
τ 3.80 (*s.*, 8 protons)
$\left.\begin{array}{l}\tau_A\ 7.52 \\ \tau_B\ 8.12\end{array}\right\}$ AB system
$J_{AB} = 11.0$ Hz.
(Room Temperature)

protons show no undue shielding (AB system, τ_A 7.44, and τ_B 8.07), and in fact resonate in a region close to that of the bridging methylenes in the 7,14-dihydro-*anti*-dimethano-[14]annulene (AB system, τ_A 7.27 and τ_B 8.54) [159].

The n.m.r. of *39* exhibits a temperature dependence [159,244b,247] which indicates that the room temperature spectrum consists of the average of two structures, *39a* and *39b*, which rapidly interconvert by fluctuation of the double bonds. At low temperatures (-135 °C in carbonyl sulphide and 20% carbon disulphide) the exchange is slowed and the spectrum

39a   *39b*

reflects the distinct environments of the olefinic protons. These peripheral protons resonate in the range τ 3.4 to 4.3; the absorptions of H(2) to H(5) constitute an AA′XX′ system which is distinct from, but overlaps, the AA′XX′ multiplet of H(9) to H(12). The absorption of the H(7) and H(14) protons remains as a singlet (τ 3.67), as their environment is not altered by interconversion of *39a* to *39b*. The bridge protons, however, no longer appear as a single averaged AB system, but as two separate AB systems, one having τ_A 7.33 and τ_B 7.71, and the other with τ_A 7.61 and τ_B 8.53. An alternative averaging process, involving collapse of one bridge to the double norcaradiene *39c* is ruled out because the bridge geminal coupling constants ($J_{A,B} = 11$ Hz) are well outside the range expected for the cyclopropane methylene group in this structure [159].

39 c

The steric interaction between the two bridging methylenes which prevented the synthesis of *syn*-1,6;8,13-dimethano-[14]annulene is expected to be overcome if the methylene groups are replaced by bridging oxygen atoms and thus the synthesis of *syn*-1,6;8,13-dioxido-[14]annulene has been achieved [160], even though the near approach of the two bridging oxygens (2.55 Å) is the shortest non-bonded oxygen-oxygen distance yet reported [161]. The n. m. r. of this compound, 38 (see Table 8), clearly reflects the delocalized nature of the peripheral 14 π-electrons. The ring protons all absorb in the range of τ 2.0 to 2.4, and the near-equality of the vicinal coupling constants ($J_{2,3} = 9.0$, $J_{3,4} = 9.2$ Hz) indicates that there is no appreciable bond alternation in the perimeter of this molecule [160]. This is substantiated by the finding that the carbon-carbon bond lengths fall within the range 1.39 ± 0.01 Å [161]. The low-field position of the proton resonances and the symmetry of the spectrum thus supports the existence of a diamagnetic ring current in the 14 π-electron perimeter. Various substituted derivatives of 38 have also been investigated [127,159].

The *syn* stereochemistry of 38 has been firmly established [127,140]. That its *anti* form is not predicted to have significant resonance stabilization (as a result of decreased *p*π overlap) no doubt accounts for the fact that it has not been isolated [140]. Moreover, at high temperature the *anti*-tetrabromo precursor 40 is dehydrobrominated to *syn*-dioxido-[14]annulene, 38 [136]. This must imply a flipping of the oxygen through the plane of the fourteen carbon perimeter, presumably via 41 and the unstable 42a to 42b (which leads to 38 under reaction conditions). Indeed, after prolonged heating of either 41 or 42b an equilibrium mixture is established, favouring the *syn* compound (75%) [159]. Such a ring

40	*41*	*42 a*	*42 b*

inversion has been shown not to occur in the methano-[10]annulene series (p. 165) [130], and the n.m.r. spectrum of *43* indicates that it is also unfavourable in the oxido-[10]annulene. Thus the two methyl groups of

43

43 absorb as distinct singlets in the n.m.r., indicating that the free energy of activation for flipping in this system must be greater than 25 kcal/mole [136].

The system possessing one oxygen bridge and a *syn* methano bridge has been synthesized [159,248] and its n.m.r. spectrum (and those of its various derivatives [159]) reflects the aromatic nature of this system. For instance the peripheral protons of *syn*-1,6-methano-8,13-oxido-[14]annulene, **39** (see Table 8), absorb in a low-field region, in the form of a singlet at τ 2.25 [H(7) and H(14)] and two superimposed AA'BB' systems [H(2) to H(5), and H(9) to H(12)] centred at approximately τ 2.5 [159,162]. The two methano protons experience the shielding of the 14 π-electron ring current, and H(16) absorbs at τ 11.39. H(15) however, absorbs at τ 9.08, and this relatively low-field shift is caused by the proximity of H(15) to the bridging oxygen. The same effect has previously been observed for the half-cage compound *44* [163]. Thus H_a is deshielded relative to H_b in this system and absorbs at τ 6.45, which is about 2.5 ppm downfield from the otherwise similar *endo* protons in

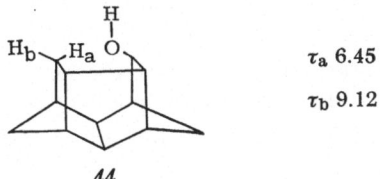

τ_a 6.45

τ_b 9.12

44

norbornane (τ 8.8). This effect, which also causes a slight shielding of H_b in *44*, and thus perhaps of H(16) in **39**, appears not to be well understood, although tentative explanations have been suggested [159,163]. The existence in **39** of a weak coupling (1.15 Hz) of bridge proton H(15) to the protons of position 2 and 5, observable in its effect on the appearance of the H(15) resonance, has been proved by double irradiation experiments. However H(7) and H(14) are not coupled to either of the bridge protons [159,162].

In the dioxido system, a bridge flipping of the dihydro precursors, $41 \leftrightarrows 42b$, was accomplished by heating at 80—100 °C for about forty hours. In the present case, synthesis of the *anti*-1,6-methano-8,13-oxido-[14]annulene has not been possible, because the *anti*-dihydro precursor, presumably as the open form *45a*, (c.f. *41*), as derived from the *anti*-tetrabromide, isomerizes spontaneously to the *syn* compound *45b* at room temperature [159]. Consequently, *syn*-1,6-methano-8,13-oxido-[14]-annulene, **39**, is the ultimate product from the *anti* precursors.

45 a   *45 b*

It was previously observed that the *syn*-dimethano-[14]annulene, *38*, is subject to a very large steric interference of the inner protons of the two bridges, and for this reason is not expected to be stable. Such a problem is identical to the interference of the two internal protons in di-*trans*-[10]annulene, *1*, which was overcome by the replacement of the protons in question with a bridging methylene group. In the present case, the application of the same principle has lead to the synthesis of 1,6;8,13-propano-[14]annulene, **40** (see Table 8) [164,165]. Such a compound allows the observation of the effect of the 14 π-electron system on the protons of bridging groups, unperturbed by the proximity of an oxygen atom. As expected the bridge protons are shielded and resonate as two triplets at τ 10.61 and τ 11.16. The assignment of these triplets to the "internal" methylene protons and the "external" methine protons, respectively, is based on studies of *46*, the di-deuterated derivative of **40**, formed from deuterium exchange in the ketone precursor [136]. The peripheral protons of **40** appear in a low-field region (singlet at τ 2.12, AA'BB' system at τ 2.35) similar to that for **38** and **39**.

46

The n.m.r. spectra of the unbridged [14]annulenes discussed earlier (4 to 9, Table 4) consist of deshielded outer protons resonating in the range τ 0.36 to 2.7, and inner protons shifted upfield to between *ca.* τ 10 to 15. The chemical shifts of the outer and "inner" (i.e. bridging) protons in the bridged 14 π analogues of both pyrene (35 to 37) and anthracene (38 to 40) perimeters fall within these ranges. Despite considerable degrees of distortion from planarity in some of these bridged and unbridged molecules, and despite large variations in structure between the three types of [14]annulenes, all reflect a common effect which is abundantly apparent from their n.m.r. spectra — namely that each compound possesses a diamagnetic ring current. The degree to which the ring current is present must largely depend on the planarity of each compound. Thus the slightly buckled [14]annulene, 9, has a smaller separation between inner and outer protons (ca. 8.4 ppm) than does the planar 1,8-didehydro-[14]annulene, 10 (ca. 14.5 ppm). Such an argument may also explain why the deshielding of the outer protons in the pyrene-perimeter bridged compounds is more pronounced than for some of the more hindered unbridged [14]annulenes, and for the less planar anthracene-perimeter systems. Thus 35 would seem to approach more nearly the behaviour expected for the planar [14]annulene itself (if it could exist). The protons incorporated into the bridges in both types of bridged compounds are strongly shielded, and, although they cannot strictly be considered as models for the inner protons in the unbridged series, they do establish the nature of the shielding effect which a diamagnetic ring current gives rise to in that region in and above the centre of the area which it circumscribes.

18π-Systems

From the early work on the n.m.r. spectra of porphyrins [166–168] it was apparent that these compounds, and their dihydro (chlorin) and tetrahydro derivatives exhibit a characteristic deshielding of the peripheral protons and large, upfield shifts of the nitrogen-bound hydrogen atoms. These effects were recognized as being due to a diamagnetic ring current [166,169] and thus this class of compounds can be considered as the [18]annulene-like, having two nitrogen atoms in the path of 18 π-electron delocalization, as in either *47a* or *47b*. The two bridging nitrogen groups are assumed not to perturb too seriously the [18]annulene character of the molecule. Furthermore, the additional double bonds in porphyrins appear to be minor perturbations, since the essential character of the n.m.r. spectra remains unchanged on progression to the di- and tetrahydro derivatives.

Extensive work in recent years has resulted in the tabulation of many such spectra, and we shall not attempt a complete coverage of this

47 a   *47 b*

area [170]. Rather, a few examples of simple systems in the series will be chosen to illustrate the general trends.

The chemical shifts in deuteriochloroform of the *meso* protons and nitrogen-bound protons in 1,2,3,4,5,6,7,8-octaethylporphine and its di-hydro and 1,2,3,4-tetrahydro derivatives are given in Table 9, (**41, 42** and **43**, respectively) [171]. In the porphine the *meso* protons (τ —0.07) resonate more than 3 ppm lower than benzene, and the imino protons (τ 13.65) are about 13 ppm higher than that of pyrrole. The presence of a ring current is also clearly implicated in **42** and **43**.

However, the use of chloroform as a solvent has certain disadvantages, in particular a concentration dependence of spectra in this solvent has been noted [172]. Furthermore, many porphyrins, including porphine itself, are only sparingly soluble in chloroform, and this has lead to wide use of trifluoroacetic acid (TFA) as a solvent, in which these compounds exist as di-cations. Protonation of two of the nitrogen atoms results in an increased separation between the chemical shifts of the internal and external protons. Thus, in deuterio-TFA the *meso* protons of the octaethylporphine, **41**, are shifted even further downfield to τ —0.98, while the imino protons, whose increased shielding more than compensates for the positive charge on the nitrogen atoms, are shifted up-field to τ 14.65 [171,173].

It is claimed that this effect reflects an increased ring current in the porphyrin di-cation, arising from the greater resonance stabilization of such a structure possessing four equivalent nitrogen atoms [169]. Alternatively, an increase in the separation of chemical shifts could arise from the deshielding effect of the two positive charges on the external protons, and an increased shielding of the inner protons by the removal of two known, (strong) deshielding factors, namely hydrogen-bonding [174] and the anisotropy of the nitrogen lone pairs [175].

Because the effect of di-protonation is seen to be simply a magnification of an effect already present in the neutral molecule, the spectrum of the porphine di-cation, **44** (see Table 9), yields information concerning porphine itself which is too insoluble in chloroform to enable its spectrum in this solvent to be observed.

Table 9*. *Porphyrins and their analogues*[c])

Compound	Chemical shifts (τ)	Solvent	Ref.
  41 1,2,3,4,5,6,7,8-Octaethyl-porphine	—0.07 (4H, s., *meso* protons) 13.65 (2H, N—H)	$CDCl_3$	[171) a)]
Di-protonated **41**	—0.98 (s., *meso* protons) 14.65 (N—H)	TFA-d_1	[171, 173)]
  42 1,2,3,4,5,6,7,8-Octaethyl-chlorin	0.32 (H_β) 1.16 (H_a) 12.49 (N—H)	$CDCl_3$	[171)]
  43 1,2,3,4,5,6,7,8-Octaethyl-1,2,3,4-tetrahydroporphine	1.51 (1H, H_γ) 2.53 (2H, H_β) 3.14 (1H, H_a)	$CDCl_3$[b)]	[171)]

* See footnote to Table 4, p. 134.

Table 9 (continued)

Compound	Chemical shifts (τ)	Solvent	Ref.
44 Di-protonated porphine	—1.22 (4H, *meso* protons) 0.08 (8H, β-protons) 14.40 (4H, N—H)	TFA	[168]
45 X = S	—0.06 (2H, *s.*) ⎱ *meso* 0.00 (2H, *s.*) ⎰ protons 0.02 (2H, *s.*, thiophene protons) 14.98 (N—H)	CDCl$_3$	[176]
46 X=O	—0.12 (2H, *s.*) ⎱ *meso* 0.02 (2H, *s.*) ⎰ protons 0.31 (2H, *s.*, furan protons)	CDCl$_3$	[176]
47 X = O Di-protonated	—1.71 (4H, *s.*, *meso* protons) —0.98 (4H, *s.*, furan protons)	TFA	[176]
48 X = S	—0.59(2H, *s.*) ⎱ *meso* —0.05(2H, *s.*) ⎰ protons —0.01(2H, *s.*) ⎱ furan and 0.31(2H, *s.*) ⎰ thiopene protons	CDCl$_3$	[176]

[a] See also Inhoffen, H. H., Fuhrhop, J-H., Voigt, H., Brockmann, H., Jr.: Lie-bigs Ann. Chem. *695*, 133 (1966).

[b] Measured at successive dilutions; final steady values quoted.

[c] See also Addendum (p. 208).

Thus in TFA the *meso* protons of porphine resonate at τ —1.22, the β-protons at τ 0.08, and the imino protons upfield at τ 14.40 [168]. That the *meso* protons[q] are more deshielded than the other peripheral protons is reflected in the ring-current calculations [169].

Some porphyrin-like aromatic macrocycles, in which one or two of the pyrrole nitrogen atoms have been replaced by oxygen or sulphur, have been studied [176], and their n.m.r. shows the same characteristic deshielding of the peripheral protons (see **45**, **46**, **47** and **48**, Table 9).

These few typical examples (**41** to **48**) indicate the dramatic effects operating in these systems, where the *meso* protons resonate about 5 ppm downfield from normal olefinic protons, and the imino protons, even when attached to a positively charged nitrogen atom, are shifted to very high fields. This behaviour is indicative of a substantial diamagnetic ring current in these molecules and lends credence to their inclusion as bridged [18]annulenes.

The porphyrins were considered as bridged [18]annulenes arising from hypothetical diaza-[18]annulenes (e.g. *48*), systems which contain two heteroatoms in the path of conjugation. We shall now consider a series of bridged compounds which more strictly adhere to the definition of an annulene, in that they are (tri)bridged derivatives of [18]annulene itself, where the six inner hydrogen atoms are replaced by three bridging groups X as in *49*, where X can be O, S or NH, or various combinations

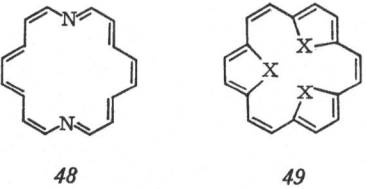

48   **49**

of these. In [18]annulene the crowding of the six inner protons is not so severe as to preclude a reasonable planarity, and a typically aromatic spectrum has been observed for this compound (**21**, see Table 4). In many of the bridged compounds of type *49*, however, the formation of the bridges is not sterically advantageous, and in some cases non-planarity, and consequent lack of delocalization occurs.

---

q) Systematic studies in TFA have shown that the *meso* protons in various octa-alkylporphyrins [168] resonate at τ —1.00$\pm$0.02, and thus the effect of alkyl substituents at all β-positions is to increase the shielding of the *meso* protons by about 0.22 ppm.

This difficulty arises especially when the bulky sulphur atom forms the bridge. Thus the n.m.r. spectrum of [18]annulene 1,4;7,10;13,16-trisulphide, *50* [177,178] (in carbon tetrachloride), shows two peaks at τ 3.32 and 3.27, arising from six identical protons in the β-positions of

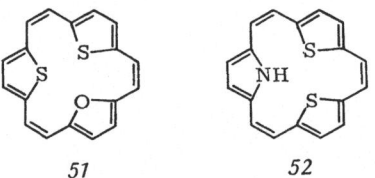

50

the thiophene rings, and from the other six identical protons, respectively. These chemical shifts reflect the lack of aromaticity which follows from the lack of planarity, [178] and which indeed was predicted [179] by a calculation on this molecule.

Similar internal crowding occurs for [18]annulene 1,4-oxide-7,10; 13,16-disulphide, *51* [180,181], and 1,4-epimino-[18]annulene 7,10;13,16-disulphide, *52* [182]. The protons of *51* resonate as four doublets and two singlets between τ 2.90 and 3.58 [181], and those of *52* are seen as two

51   *52*

doublets and two singlets in the region τ 3.00 to 3.60, with the broad imino proton singlet located at τ 1.77 [182]. The relatively high-field absorption of the peripheral protons in these three compounds is consistent with the lack of a macrocyclic ring current in each case, a conclusion which is corroborated for *52* by the low-field resonance of its imino proton. The ultraviolet and visible spectra of the three compounds are similar, emphasizing their inter-relation. These three systems, then, may be considered to consist of three heterocyclic rings joined by three olefinic linkages.

When two more of the bridging groups are the less bulky oxygen atom, near-planarity results, and the concomitant effect on the delocalization of the 18 π-electrons is seen in the n.m.r. spectra of [18]annulene 1,4; 7,10-dioxide-13,16-sulphide [183,184] and [18]annulene 1,4;7,10;13,16-trioxide [180,185,186] (49 and 50, respectively, see Table 10). The spectrum of 49 consists of two AB quartets and two singlets in the range τ 1.00 to 1.72 [184], while that of 50 shows two peaks of equal area at τ 1.32 and 1.34 [185]. These chemical shifts, which are in the same region as those of the outer protons of [18]annulene, represent a downfield shift from the resonances of acyclic vinyl furans (τ 3.0 to 3.9) [187], and are quite distinct from those of the three non-planar bridged compounds discussed earlier. The presence of an appreciable diamagnetic ring current in 49 and 50 is thereby implied; in addition, the u.v. spectra of these two compounds are similar to that of the isoelectronic tridehydro-[18]annulenes (18, 19), suggesting that the π-electrons of the three bridging heteroatoms are not critically involved in the peripheral conjugation [184,185].

Although 50 possesses a ring current, the derivative 53, containing three methoxycarbonyl substituents does not. This is a result of a severe steric interaction between the substituents and the furan rings, which is evidently sufficient to destroy the planarity of the macrocyclic ring [185].

53

In the series [18]annulene trisulphide, oxide-disulphide, dioxide-sulphide and trioxide, the relationship between planarity and delocalization has again been demonstrated. The trisulphide and oxide-disulphide are predicted from Catalin models to be non-planar and their n.m.r. does reflect the absence of a ring current over the total 18 π-electron system. Models suggest that the dioxide-sulphide is near-planar and the trioxide planar, and indeed these compounds are aromatic [183,185].

Two bridged [18]annulenes of type 54, possessing four bridging groups (X = O, N, or NH) have also been investigated [188]. Thus the two furan-pyrrole macrocycles 51 and 52 (see Table 10) possess *meso* protons which are very strongly deshielded (τ 0.51 to 0.82), in analogy

185

Table 10*. *Bridged [18]annulenes*

Compound	Chemical shifts (τ)	Solvent	Ref.
49 [18]Annulene-1,4; 7,10-dioxide-13,16-sulphide	1.62 ⎱ (H$_2$, H$_3$, H$_8$, H$_9$) 1.70 ⎰ $J_{2,3} = J_{8,7} = 2.4$ Hz[a]) 1.64 (s., H$_{14}$, H$_{15}$) 1.13 ⎱ 1.59 ⎰ (H$_{11}$, H$_{12}$, H$_{17}$, H$_{18}$) $J_{11,12} = J_{17,18} = 12.5$ Hz 1.15 (s., H$_5$, H$_6$)	CDCl$_3$	184)
50 [18]Annulene-1,4;7,10;13,16-trioxide	1.32 1.34 (equal area)	CCl$_4$	185)
51	0.75 (2H, s.) ⎱ *meso* 0.82 (1H, s.) ⎰ protons	CDCl$_3$	188)
52	0.51 (1H, s.) ⎱ 0.67 (1H, s.) ⎰ *meso* 0.76 (1H, s.) ⎰ protons	CDCl$_3$	188)

* See footnote to Table 4, p. 134.
a) The coupling constant of 3.4 Hz quoted in ref. 184 is not consistent with the chemical shift data therein.

with those of the porphyrins. In this case reasonable planarity is not prevented by small oxygen and nitrogen bridging groups, and the aromaticity of these systems is evident.

54

22π- to 36π-Systems

It was noted earlier that no example of an unbridged [22]annulene had been reported. For this reason the syntheses [176,188] of two [22]annulenes bridged by five heterocyclic groups (**53** and **54**, see Table 11) are particularly significant. [22]Annulenes are predicted[2c] to prefer a delocalized structure, without the onset of severe bond alternation [7]. The n.m.r. spectrum of mono-protonated **53** is clearly indicative of the presence of a diamagnetic ring current, the *meso* protons resonating at a very low field (τ —2.28) and the NH protons being greatly shielded (τ 15.5 and 16.8). By analogy with the porphyrins, such protonation of a bridging nitrogen, while altering the magnitude of the chemical shifts in the periphery to some extent, is not expected to affect the conclusion regarding the aromaticity in this system. The spectrum of the free base **54** lends support to this thesis.

Two isomeric, bridged [24]annulenes, **55** and **56** (see Table 12), have been prepared [186,189] and they, like their unbridged analogues, appear to sustain a (small) paramagnetic ring current. The most likely configurations of the [24]annulene 1,4;7,10;13,16;19,22-tetraoxides are as shown, the configuration having four *trans* double bonds linking the furan segments being assigned to **56** largely on the n.m.r. data, as three singlets only (1:2:1) are observed. These configurations are supported by infrared studies, and alternative stereochemistries are considered less likely [189]. In both isomers there are two distinct regions of absorption; a high-field (τ 4.1 to 5.2, outer protons) and a low-field (τ 1.3 to 1.4) region. The low-field resonances correspond to three protons for **55** and four for **56**, and are due to the inner protons, which, like those inner protons in the unbridged [24]annulenes **25** and **26**, are deshielded by the paramagnetic ring current in the peripheral system. The electronic spectra of these molecules are very similar to those of [24]annulene and tetradehydro-[24]annulene, confirming that there is a related peripheral delocalization in each system.

Table 11*. *Bridged [22]annulenes*

Compound	Chemical shifts (τ)	Solvent	Ref.
53 (monoprotonated)	—2.28 (4H, *s.*, *meso* protons) 15.5, 16.8 (NH)	CDCl₃-TFA (4:1)	188)
54	—0.52 (1H, *t.*) } *meso* —0.06 (2H, *s.*) } protons 0.42 (4H, furan protons) 14.85 (2H, NH)	CDCl₃	176)

* See footnote to Table 4, p. 134.

Members of the [28]annulene series have yet to be reported, and no bridged analogues have been synthesized to date.

[30]Annulene pentoxide, *55* (no configuration implied), has been isolated as two geometric isomers, the configurations of which could not be determined 186,189). The n.m.r. spectra 189) of both isomers (practically invariant from —60 °C to +100 °C) showed absorptions (τ 3.3 to 4.0) similar to those of the acyclic vinyl furans 187) and of the unbridged pentadehydro- and tridehydro-[30]annulenes110,116). Once again the ultraviolet spectra of these compounds are closely related to the unbridged analogues. The peripheral atoms in these two isomer of *55* do not appear to deviate markedly from coplanarity 189), and thus the apparent absence of ring currents must reflect severe bond alternation.

Table 12*. *Bridged [24]annulenes*

Compound	Chemical shifts (τ)		Ref.
			189)
55	Inner protons 1.3—1.4 (3H)	Outer protons 4.1—5.2 (13H)	
			189)
56	Inner protons 1.29 (4H, *s.*)	Outer protons 4.1—5.2 (12H)	

* See footnote to Table 4, p. 134.

The spectrum of the $4n\pi$ system [36]annulene hexoxide [189], *56*, also reflects the absence of a ring current, as the protons absorb in the range τ 2.4 to 4.2.

55　　　　　　*56*

The prediction [2c], then, that in the $(4n+2)\pi$ series the [22] and lower annulenes should be aromatic while the [26] and higher annulenes should be non-aromatic is supported by the evidence available to date. However, the investigation of unbridged [22]annulenes would add weight to this conclusion. Of even more importance would be the availability of further data on [26]annulene (at the present time only a tridehydro-[26]annulene, 27, has been studied, see Table 4).

Thus, from a consideration of both bridged and unbridged annulenes ranging from 6 to 36 peripheral π-electrons, it has been seen that, in planar systems, n.m.r. spectroscopy can distinguish clearly between those aromatic $(4n+2)\pi$ systems and the anti-aromatic $4n\pi$ cases, until, at very larger ring size where Hückel's rule becomes invalid, both series merge into non-aromatic behaviour.

2. Charged Annulenes

For ease of presentation, it is convenient to divide those ions which can be considered as "charged annulenes" into two sub-groups. The first sub-group ("Type 1"; see Fig. 6) comprises those systems possessing an odd number of carbon atoms in the peripheral ring system. Obviously no closed shell neutral annulene with such a carbon skeleton is possible. These systems bear a formal single positive or negative charge, and will include both bridged and unbridged molecules.

With one exception, all of the compounds in the sub-group "Type 2" are formally derived from neutral, anti-aromatic annulenes by the addition or removal of two electrons from the π-system (see Fig. 6). It has been pointed out earlier that the occurence of bond alternation and paramagnetism in the anti-aromatic annulenes is due to the near degeneracy of the non-bonding orbitals of these systems. This behaviour may be reversed by either removing the two non-bonding electrons (di-cation formation), or adding two additional electrons (di-anion formation), thus giving a closed shell configuration for which no Jahn-Teller distortion is predicted (see Fig. 6).

Our interest is mainly focused on ring current effects in these quasi-annulenes. However, it is well known that deviations from electroneutrality at a carbon atom affects the shielding at that carbon and at the nearby protons. This latter effect is electrostatic in origin and is determined principally by the component of the electric field in the direction of the C—H bond. It is necessary to separate out the shielding contributions of such electrostatic effects, in order to make a meaningful comparison between the ring current effects in the ions and the neutral annulenes. Such a separation, however, is fraught with difficulties. In particular, the relationship between charge density and proton chemical shift has in the main been built up empirically from such systems as the ions we are about to consider by subtracting all contributions to the chemical shifts other than the charges [190-192]! A further problem arises in that some of the ions, unlike the neutral annulenes, embody quite severe angle distortions. This will undoubtedly influence the hybridization of the carbon atoms and may alter the proton shieldings because of the concomitant differences in effective electronegativity which result.

Nevertheless, a generally accepted approximation is that a proton is shielded by about 10 ppm by the presence of one electronic charge on the sp^2 hybridized carbon atom to which it is bound, and similarly a unit positive charge produces a deshielding of 10 ppm [193,194]. This value will be used in this review to aid comparison with the uncharged species, but too much weight should not be placed on the shifts thus obtained, particularly in view of the unknown contribution of solvent and counterion.

Where possible, only the parent system will be discussed, exceptions being those cases where the unsubstituted compound has not yet been reported.

Type 1

The cyclopropenyl cation [195,196] (**57**, see Table 13), a $2\pi3C$ system [r], has no analogy in the neutral annulenes, but represents a lowest member in the series of cyclic $(4n+2)\pi$-electron systems $(n=0)$. The n.m.r. spectrum of cyclopropenyl hexachloroantimonate shows a singlet at τ —1.1 [195]. The deshielding of a single positive charge distributed over three carbons is taken as approximately 3.3 ppm, giving rise to a "corrected" chemical shift of τ 2.2 for this system. Although no direct comparison to a neutral molecule is possible, this low-field resonance (downfield from benzene) is consistent with the formulation of an aromatic molecule possessing a conjugated 2 π-electron system.

It has been found that ^{13}C chemical shifts are quite sensitive to local π-electron densities, with a proportionality constant of ± 160 ppm (with reference to benzene) per $\mp$ charge, holding to good accuracy. Such shifts, then, provide a useful experimental criterion for the charge densities in a particular molecule (the effect of ring current on ^{13}C shifts is probably negligible). In the present case the ^{13}C chemical shift of

---

[r] For nomenclature used see Winstein, S.: Aromaticity, Special Publication No. 21, The Chemical Society, London, 1967, p. 5.

Table 13*. *Type 1 ions*

Compound	Chemical shift (τ)	τ Corrected for charge	Solvent	Counter-ion	Conc. temp.	Ref.
57 2π3 C	—1.1 $J_{13C-H} = 265$ Hz	2.2	CH_3CN with $SbCl_5$	$SbCl_6^-$		195)
			CH_3CN	BF_4^-		
			SO_2	BF_4^-		
	—0.87	2.4	FSO_3H	$SbCl_6^-$		
	—0.80	2.5	FSO_3H	FSO_3^-		196)
			$ClSO_3H$	$ClSO_3^-$		
	$\delta^{13}C = +17.8$ ppm from CS_2. $J_{13C-H} = 262$ Hz		SbF_5- SO_2	SbF_5Cl^-	—60 °C	206)
58 6π5 C	4.70 a)	2.7	DMSO	Na^+	2%, 30 °C	198)
	4.52 a)	2.5	CH_3CN	Na^+	2%, 30 °C	
	4.45 a)	2.5	THF	Na^+	2%, 30 °C	
	4.47 a)	2.5	THF	Li^+	2%, 30 °C	
	4.44 a)	2.4	THF	Na^+	8 mole %	190)
	4.40 a)	2.4	THF	Na^+	4	
	4.36 a)	2.4	THF	Na^+	2	
	4.33 a)	2.3	THF	Na^+	1	
	4.32 a)	2.3	THF	Na^+	0.5	
	4.31 a)	2.3	THF	Na^+	0 (extrap.)	
	4.56 a)	2.6	CH_3CN	Li^+	4	
	4.56 a)	2.6	CH_3CN	Li^+	2	
	4.56 a)	2.6	CH_3CN	Li^+	1	
	$\delta^{13}C = +25.7$ ppm from benzene. (calc. +32 ppm). $J_{13C-H} = 157$ Hz		THF			192)
59 6π7 C	0.79 a)	2.2	DMSO	ClO_4^-	2%, 30 °C	198)
	0.86 a)	2.3	CH_3CN	ClO_4^-	2%, 30 °C	
	0.80 a)	2.2	DMSO	BF_4^-	2%, 30 °C	
	0.86 a)	2.3	CH_3CN	BF_4^-	2%, 30 °C	
	0.82 a)	2.3	CH_3CN	Br^-	0.2 mole %	190)
	$\delta^{13}C = —27.6$ ppm from benzene. (calc. —22.9 ppm) $J_{13C-H} = 171$ Hz		$H_2O/$ HBF_4	BF_4^-		192)

* See footnote to Table 4, p. 134.

Table 13 (continued)

Compound	Chemical shift (τ)	τ Corrected for charge	Solvent	Counter-ion	Conc. temp.	Ref.
60 10π9 C	2.96 3.15	1.9 2.0	THF-d_8 THF-d_8	K$^+$ Li$^+$		201)
	3.28 3.18	2.2 2.1	THF DMSO-d_6	Li$^+$ N(Et)$_4^+$		202)
	$\delta^{13}C = +19.0$ ppm from benzene. (calc. $+17.8$) $J_{13C-H} = 137$ Hz			Li$^+$		
61 10π11 C	0.4—1.7 (*m.*) 10.3, 11.8 (bridge protons: $J = 10$ Hz)	1.3—2.6	CD$_3$CN	BF$_4^-$		203)
62 10π9 C	3.2 (2H, *d.d.*) 4.0—4.6 (5H, *m.*) 10.7, 11.2 (bridge protons: $J = 7.5$ Hz)	2.1 2.9—3.5	DMSO-d_6	Na$^+$		204a)
	2.94 (2H, *d.d.*) 3.98 (5H, *m.*) 10.45, 10.95 (bridge protons: $J = 7.5$ Hz)	1.8 2.9	DMSO-d_6	Na$^+$		204b)
63 18π17 C	—0.47 (*t.*, H$_{15}$) 0.09 (*q.*)⎱ H$_6$, 0.18 (*q.*)⎰ H$_{13}$ 0.34 (*d.*, H$_4$, H$_{17}$) 1.60 (*d.*)⎱ H$_7$, 2.16 (*d.*)⎰ H$_{12}$ 5.38 (*s.*, OCH$_3$) 18.54 (*t.*)⎱ H$_5$, 18.85 (*t.*)⎰ H$_{14}$, 19.09 (*t.*)⎰ H$_{16}$	Outer protons: —1.1 to 1.6 Inner protons: 17.9 to 18.5	THF-d_8	Li$^+$ or K$^+$	—35 °C	205)

For compound **63**:
$J_{5,4} = J_{5,6} = J_{13,14} = J_{14,15} = J_{15,16} = J_{16,17} = 13$ Hz
$J_{6,7} = J_{12,13} = 9$ Hz

a) Chemical shifts originally quoted relative to benzene. The conversion to the τ-scale is made by adding 2.73 ppm to the quoted values.

+17.8 ppm from $^{13}CS_2$ (47.8 ppm downfield from benzene) is in reasonable agreement with that calculated (+12.3 ppm, or 53.3 ppm from benzene) on the basis of complete delocalization of 2 π-electrons over this positively charged system [197].

The cyclopentadienyl anion (6π5C) and the tropylium cation (6π7C) are planar, aromatic ions directly analogous to benzene. The n.m.r. spectrum of the cyclopentadienyl anion has been extensively studied in various solvents and at various concentrations, with both sodium and lithium as the counterion [190,198,199] (58, see Table 13). The proton resonance varies within the range τ 4.3 to 4.7, the variability reflecting, in the main, the changing conditions, although results by different workers on similar solutions have not always been compatible [190,198]. The correction for the negative charge yields a shift of about τ 2.3 to 2.7, which is clearly in an aromatic region.

The tropylium cation has also been studied under various conditions [190,198,199], with chemical shift values of τ 0.79 to 0.86 being reported (59, see Table 13). The corrected chemical shift (τ 2.2 to 2.3) is slightly lower than the range for the cyclopentadienyl anion, in agreement with the prediction [200] that the ring current effect should be larger in the tropylium ion than in C_6H_6 or $C_5H_5^-$, due in part to the larger area of the carbocyclic ring. However, the accuracy of our corrected values is such that small differences in chemical shift cannot be deemed significant.

The ^{13}C chemical shifts for these two systems have also been determined [192], and compare quite favourably with the calculated values (see Table 13).

A charged analogue of the unknown [10]annulene, the cyclononatetraenyl anion (10π9C) possesses aromatic character as demonstrated by its ^1H and ^{13}C n.m.r. spectra [201,202] (60, see Table 13). The protons of this compound absorb as a sharp singlet in an aromatic region (τ 2.96 to 3.28), the exact position being dependent on counterion [201,202] and concentration [202]. The correction for the shielding effect of the electronic charge distributed over nine carbon atoms further enhances the low-field position of these protons (τ 1.9 to 2.2). The ^{13}C spectrum [202] (doublet at 19.0 ppm upfield from benzene) is also entirely compatible with a delocalized aromatic compound, the predicted chemical shift of such a system being 17.8 ppm.

Two bridged, ionic 10 π systems directly analogous to the bridged 1,6-methano-[10]annulene discussed earlier, have been studied. Thus the cation 61 and anion 62 (see Table 13) bear a similar relationship to 1,6-methano-[10]annulene, 28, as do the tropylium ion and cyclopentadienyl anion to benzene. The n.m.r. spectrum of cation 61 [17,203] is characterized by low-field absorptions of the peripheral protons (τ 0.4 to 1.7) and by enhanced shielding of the bridging protons (τ 10.3, 11.8) as was observed for the neutral compound. The anion **62** [17,204)] has absorptions near τ 3, and τ 4, and again the high-field shift of the bridging protons (τ 10.4 to 11.2), together with their large geminal coupling (7.5 Hz) which was present in **61** and **28** (10 Hz and 6.9 Hz, respectively), and which is compatible with the structures shown.

The cation possesses a more favourable geometry than 1,6-methano-[10]annulene, being able to incorporate the 10 π-system into a more planar structure. It is evidently more stable than the tropylium ion, as it can be formed from its neutral hydrocarbon precursor *57* by a hydride-transfer reaction with tropylium tetrafluoroborate [136)]. The anion, on the other hand, involves substantial steric strain and is not as planar as either the cation or the neutral bridged [10]annulene. Whereas formation of the cation from *57* was seen to be most favourable, the anion precursor *58* is much less acidic than cyclopentadiene, although in the presence of

57   *58*

the dimethylsulphoxide anion, **62** is readily formed. The chemical shifts of the peripheral protons in **61** and **62**, when corrected for the charge distributed over eleven and nine carbons respectively, yield values of τ 1.3 to 2.6 for the cation, and τ 1.8 to 3.5 for the anion, comparing well with the shifts (τ 2.7 to 3.1) of the aromatic neutral molecule (note, however, that the equality of charge densities in the carbocyclic perimeter is no longer dictated by symmetry).

The increase in chemical shift on progression from the cationic to neutral to anionic species (in so far as it is significant) parallels the decreasing ring area and the increasing departure from planarity in this series. Furthermore, the corrected shifts (τ 1.9 to 2.2) of the cyclonona-tetraenyl anion, **60**, in which steric inhibition to planarity is minimal, are comparable to those of the bridged cation. In any case, the n.m.r. spectra of the two bridged [10]annulene-like ions clearly reflect their aromatic nature.

Among the higher annulenes, a mono-charged analogue has been reported in one case, that of the seventeen-carbon macrocycle **63** (18 π 17C) (see Table 13) [205)]. This anion is a counterpart of the tridehydro-[18]annulenes, and contains 18 out-of-plane π-electrons. Like the neutral higher annulenes, it possesses both inner and outer protons, the resonances of which provide a dramatic proof of aromaticity. The outer protons resonate over the range τ —0.47 to 2.16, and the inner protons appear as three superimposed triplets at remarkably high field (τ 18.54, 18.85 and 19.09) [205]. The assumption of equal charge densities at each of the peripheral carbon atoms, which has been used in evaluating the charge correction factor, is not expected to be a very good approximation in this case, due, in part, to the presence of the acetylenic bonds. Indeed, the wide spread in chemical shifts observed for the outer proton resonances does indicate an uneven charge distribution. In any case, the theoretical correction of —0.6 ppm is trivial in view of the large chemical shift differences which provide such strong evidence for aromaticity and the presence of a diamagnetic ring current in this system.

Type 2

The cyclobutenium di-cation, *59*, is formally derived from the (as yet unisolated) anti-aromatic cyclobutadiene by the removal of two π-electrons. This di-cation, like the cyclopropenyl cation, represents a member of the aromatic sub-group possessing just two π-electrons. The tetramethyl derivative of *59*, **64** (see Table 14), has been studied [206], and its n.m.r. spectrum appears to possess the expected aromatic characteristics. The methyl groups resonate at τ 6.32 in SbF_5—SO_2 solutions and at τ 5.89 in SbF_5—SO_2ClF. A reasonable model for this

$$\tau_1 = \tau_3 = 7.19$$
$$\tau_2 = 7.85$$

59   *60*

compound is *60* (the 1,2,3,4,4,5-hexamethylcyclopentenyl cation) [207], in which the 1- and 3-methyl groups (τ 7.19) are similarly attached to sp^2-carbon atoms which have a charge density of approximately 0.5. The relative deshielding of the protons of **64**, then, probably reflects a ring current effect. The ^{13}C chemical shifts of **64** constitute further proof of its delocalized nature, as the observed value [206] for the ring carbons of 80 ppm downfield from benzene is exactly that calculated for this system.

196

It was seen earlier that cyclooctatetraene, *3*, is a non-planar, non-aromatic compound. The addition of 2 π-electron converts this system to the di-anion **65**, and the new species appears to be planar and highly resonance stabilized. The protons of this 10π8C system resonate at τ 4.3 [190,208] (there is a discrepancy in some early work [209]), the charge-corrected value being τ 1.8, which is clearly in an aromatic region. The ^{13}C chemical shift [192] of +42.5 ppm from benzene is also as expected, the calculated value being +40 ppm .

A similar dramatic change is observed when the three *trans*-15,16-dialkyldihydropyrenes, *61*, discussed earlier are converted to their respective di-anions, **66, 67** and **68** (see Table 14) [157]. In this case, the conversion corresponds to a change from the aromatic 14 π-electron molecule to

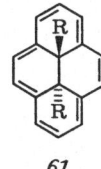

61

R = methyl, ethyl
or *n*-propyl

the anti-aromatic 4*n* system containing 16 π-electrons, and is clearly marked by the change in the proton chemical shifts [157]. Whereas the external protons in the neutral precursors (**35, 36, 37**, see Table 8) were deshielded and resonate in the range τ 1.33 to 2.05, they are now observed in the range τ 12.50 to 13.96. Furthermore the interior protons at the α-carbon atoms, which were found at τ 13.95 to 14.25 in the neutral molecules are shifted to τ −11.0 to −11.24 in the di-anions, a downfield shift of 25 ppm! The protons at the β-carbon in the charged species are also deshielded but to a lesser extent (τ −0.70 to −2.59), and those at the γ-carbon atoms in the di-*n*-propyl di-anion resonate at τ 4.49. This change in the proton shielding pattern on formation of the di-anion thus gives unassailable proof of the concomitant change from a diamagnetic to a paramagnetic ring current.

A contrasting conversion from an anti-aromatic to an aromatic system is illustrated by the addition of 2 π-electrons to [16]annulene, and the results are equally remarkable. It was seen that in [16]annulene itself a dynamic equilibrium exists between the two configurations *16a*

Table 14*. *Type 2 ions*

Compound	Chemical shift (τ)	τ Corrected for charge	Solvent	Counter-ion	Conc. temp.	Ref.
Me Me [2+] Me Me **64** 2π4 C	6.32		SbF_5—SO_2	2 Cl⁻	—65 °C	206)
	5.89		SbF_5—SO_2ClF	2 Cl⁻	—78 °C	
	$\delta^{13}C = -14.4$ ppm from CS_2 or $= -80$ ppm from benzene.		SbF_5—SO_2		—70 °C	
	$\delta^{13}C = -15.7$ ppm from CS_2. $J_{13CCH} = 4.0$ Hz		SbF_5—SO_2ClF		—70 °C	
[2−] **65** 10π8 C	4.3 a)	1.8	THF	2 K⁺	0.6 M	208)
	4.33 b)	1.8	CH_3CN	2 Na⁺	3 mole %	190)
	4.32 b)	1.8	THF	2 Na⁺	3 mole %	
	$\delta^{13}C = +42.5$ ppm from benzene. $J_{13C-H} = 145$ Hz		THF			192)
Me [2−] Me **66** 16π14 C	—11.0 (α—CH_3) 13.19—13.96 (external protons)	11.8—12.6 (external protons)	THF-d_8	2 K⁺	—65 °C	157)

* See footnote to Table 4, p. 134.

Table 14 (continued)

Compound	Chemical shift (τ)	τ Corrected for charge	Solvent	Counter-ion	Conc. temp.	Ref.
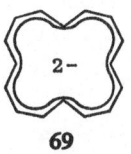**67** $16\pi 14\,C$	—11.15 (α—CH$_2$) — 0.70 (β—CH$_3$) 12.50—13.14 (external protons)	 11.1—11.7 (external protons)	THF-d_8	2 K$^+$	—65 °C	157)
68 $16\pi 14\,C$	—11.24 (α—CH$_2$) — 2.59 (β—CH$_2$) 4.49 (γ—CH$_3$) 12.56—13.14 (external protons)	 11.2—11.7 (external protons)	THF-d_8	2 K$^+$	—65 °C	157)
69 $18\pi 16\,C$	1.17 (8H) 2.55 (4H) 18.17 (4H, internal protons)	—0.1 1.3 16.9	THF-d_8	2 Li$^+$ 2 Na$^+$ or 2 K$^+$	Un- changed from —100 °C to +140 °C	210)

a) Reported as shifts relative to the lower solvent band.
b) Reported relative to benzene.

199

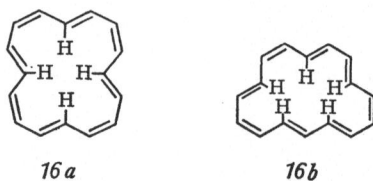

16a   *16b*

and *16b*, *16a* being favoured. The spectrum of the di-anion [210] **69** which results after prolonged contact (3 days at 0 °C) between a [16]annulene solution in tetrahydrofuran and potassium (or sodium or lithium) metal, is attributed to the di-anion having solely the *16a* type configuration, whereby the four inner protons (τ 18.17) and the two sets of different outer protons (τ 1.17, 8H and τ 2.55, 4H) uniquely arise (see Table 14). This configuration is most favourable from a resonance point of view whereas the other likely alternative *16b* possesses an increased hindrance to planarity due to the steric interference of the five internal protons [210].

Thus, by the mere addition of two electrons to [16]annulene the inner protons are shifted from τ —0.61 (—140 °C) to τ 18.17, whilst the outer protons τ 4.70 (—140 °C) are shifted downfield to τ 1.17 and τ 2.55, such noticeable shifts paralleling the change from a system possessing a paramagnetic ring current to one with a diamagnetic current.

The spectrum of the di-anion is unchanged from —100 °C to +140 °C, unlike the isoelectronic [18]annulene, and the [16]annulene precursor for both of which, at elevated temperatures, the inner and outer protons are rendered magnetically equivalent by an exchange process. It follows that if such an exchange process is possible for the di-anion system, its free energy of activation must be at least 24 kcal/mole, or some 10 kcal/mole greater than the corresponding value (13.8 kcal/mole at 140 °C) for [18]annulene. This implies that resonance plays a larger role in the di-anion than in the isoelectronic, neutral [18]annulene. Such behaviour is not hard to understand, as any bond rotation process in the di-anion will lead to a transition state involving an (energetically unfavourable) increase in the localization of electronic change [s].

---

[s] This is somewhat similar to the phenomenon observed in **63**, the mono-anion analogue of tridehydro-[18]annulene. The spread in the mono-anion (**63**) chemical shifts (between inner and outer protons) is *ca.* 16 ppm, which is to be compared with the corresponding neutral tridehydro-[18]annulenes (**18** and **19**) whose chemical shifts span about 7 ppm. This argues for a larger ring current in the charged species which in the present case would be expected to reflect a decrease in bond alternation. Again this behaviour occurs because a more completely delocalized structure is favoured by the exigencies of charge dispersal over the carbocyclic ring. Related effects may be operative in the porphyrins (the ring currents of which apparently increase on protonation) and the homoaromatics (where all well authenticated examples bear a formal charge).

3. Homoaromatic Systems

Some examples of homoaromatic systems are summarized in Table 15. The pentamethylcyclobutenyl cation [211] possesses a strong 1,3 interaction, aided by the proximity inherent in a 4 membered ring, and it thus can be considered as a 2π monohomocyclopropenyl species [212] (70, Table 15). Although the best evidence for this enhanced 1,3 π-overlap comes from ultraviolet spectroscopic studies, the n.m.r. spectrum [210] is consistent with this hypothesis. A comparison with the resonances of the methyl protons of the model compound 60 (1,2,3,4,4,5-hexamethylcyclopentenyl cation [207]) indicates that there resides a lower positive charge at the 1,3 positions and a higher positive charge at the 2 positions of the cyclobutenyl cation, thus implying partial bond character between the 1 and 3 carbon atoms, a conclusion also supported by charge density calculations based on the chemical shifts of 70 [211].

An example of a bishomocyclopropenyl system is the 7-norbornenyl cation, 71 (see Table 15) [212–214], whose chemical shifts and coupling constants are not entirely incompatible with the formulation in which C(7) interacts equally with C(2) and C(3). However, the high chemical shift of the proton on the bridging carbon atom (H(7), τ 6.76), although explained in terms of the tendency of the bridging carbon atom to rehybridize from sp^2 to sp^3 [212,213], suggests that ring current effects are not operative in this system.

In the bicyclo[3.2.1]octadienyl anion, formulated as a bishomocyclopentadienyl anion (72, see Table 15), those protons which would be involved in the cyclopentadienyl unit have resonances ranging from τ 4.61 to 7.16 [215]. Even when corrected for charge (on the naive assumption of equal charge density at all five carbon atoms) this range (τ 2.6 to 5.2) is too far upfield for protons attached to a delocalized 6π system in which a diamagnetic ring current is present, although the relative order of chemical shifts in the five protons concerned is as expected from simple HMO calculations [212,215].

The nature of the 1,3 orbital overlap in homoaromatic systems is intermediate between σ and π. Apparently the hybridization of the atoms concerned is important in determining whether a ring current effect is operative, or at least readily recognizable from the n.m.r. spectrum. Thus in systems 71 and 72 the n.m.r. spectra do not unambiguously reflect a ring current. The data is more convincing in the case of the bicyclo[5.1.0]octadienyl cation, 73 (see Table 15). The n.m.r. spectrum of this ion [216,217] is compatible only with the homotropylium structure 64.

The structure 62 is eliminated in that H(1) and H(7) resonate at about τ 3.5, which is far too low for a typical cyclopropyl proton. Furthermore, the coupling constants of H(1) to H(a) and H(b) are not as expected

Table 15*. *Homoaromatic systems*

Compound	Chemical shifts (τ)	Coupling constants (Hz)	Solvent	Counterion	Conc. temp.	Ref.
Me Me Me Me Me Me **70** $2\pi 3\,C$	7.36 (6H, C_1 and C_3) 7.63 (3H, C_2) 8.40 (6H, C_4)		CH_2Cl_2	$AlCl_4^-$	R.T. to 80 °C	211)
H_c H_e— H_b H_a H_d **71** $2\pi 3\,C$	2.93a) (H_a) 5.76a) (H_b) 6.76a) (H_c) 8.13a) (H_d) 7.56a) (H_e)	$J_{a,c}=2.5$ $J_{b,c}=2.6$ $J_{c,d}=0.8$	SO_2- SbF_5 —FSO_3H	FSO_3^-	—60 °C to —35 °C	212, 213)
	2.81b) (H_a) 5.73b) (H_b) 6.73b) (Hc) 8.16b) (H_d) 7.61b) (H_e)		SO_2- FSO_3H	FSO_3^-	—60 °C	214)
H_a 8 H_b 1 2 7 3 6 4 **72** $6\pi 5\,C$	7.55 (H_1, H_5) 7.16 (H_2, H_4) 4.61 (H_3) 6.33 (H_6, H_7) 9.13 (H_{8a}) 9.58 (H_{8b})	$J_{8a,8b}=8$ $J_{2,3}=6.7$ $J_{1,2}=5.3$ $J_{1,8a}=4$ $J_{1,8b}<2$ $J_{1,7}<2$	THF-d_8	K^+		215)
	7.51 (H_1, H_5) 7.18 (H_2, H_4) 4.65 (H_3) 9.16 (H_{8a}) 9.58 (H_{8b})		DME	K^+		

* See footnote to Table 4, p. 134.

Table 15 (continued)

Compound	Chemical shifts (τ)	Coupling constants (Hz)	Solvent	Counterion	Conc. temp.	Ref.
H_a 8 H_b (structure, 73, $6\pi 7\,C$)	1.4 (H_2 to H_6) 3.4 (H_1, H_7) 4.8 (H_{8b}) 10.6 (H_{8a})		98% H_2SO_4	HSO_4^-		[216]
	1.53 (H_2 to H_6) 3.58 (H_1, H_7) 4.90 (H_{8b}) 10.67 (H_{8a})	$J_{8a,1} = 9.8$ $J_{8b,1} = 7.5$ $J_{8a,8b} = 6.5$	conc. H_2SO_4	HSO_4^-		[217, 212]
H_a 9 H_b (structure, 74, $10\pi 8\,C$)	ca. 4.5 (H_2 to H_7) 6.0 (H_1, H_8) 10.0 (H_{9a}) [c]		THF-d_8	K^+	—60 °C	[218]
	ca. 4.8 (H_2 to H_7) 6.1 (H_1, H_8) 8.0 (H_{9b}) 10.0 (H_{9a})		DME-d_{10}	K^+	—60 °C	
H_a 11 H_b (structure, CH_2SOCH_3, 75, $10\pi 9\,C$)	3.19 (H_5, H_7) 4.18 (H_3, H_9) 5.09 (H_4, H_8) 5.28 (H_2, H_{10}) 9.62 (H_{11a} or H_{11b}) 9.90 (H_{11a} or H_{11b})	$J_{2,3} = 11.0$ $J_{3,4} = 8.3$ $J_{4,5} = 7.9$ $J_{11a,11b} = 10.5$	DMSO	Na^+		[219]

[a] Measured relative to CH_2Cl_2.
[b] Measured relative to $(CH_3)_4N^+Cl^-$.
[c] The absorption due to proton H_{9b} is obscured by solvent.

for *62* [217)], and the geminal coupling constant $J_{8a, 8b} = 6.5$ is substantially larger than that expected for cyclopropane methylene groups. The planar *63* is rejected because H(a) and H(b) are not magnetically equivalent. Thus, the low-field resonances of the protons at postions 1 to 7

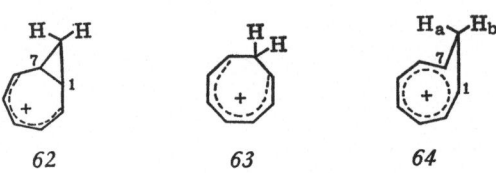

| 62 | 63 | 64 |

(τ 1.4 to 3.4) [216)], and the dramatic chemical shifts of the methylene protons (τ 10.6 for H(a), the proton above the ring, and τ 4.8 for H(b)) argue strongly for the homoaromatic 6 π system in which a diamagnetic ring current is present.

Similarly, the monohomocyclooctatetraene di-anion, **74**, (see Table 15) [212,218)] has deshielded peripheral protons, and a differential shielding of the methylene protons, whereby that proton situated above the ring (H(a), τ 10.0) is more shielded than the "outer" proton (H(b), τ 8.0). The chemical shifts of the peripheral protons (centered at *ca.* τ 4.5 to 4.8) are shifted upfield from those of the parent hydrocarbon by only 0.4 to 0.7 ppm. As the upfield shift expected from charge density considerations alone is 2.5 ppm, a deshielding effect associated with a diamagnetic current is clearly implicated [218)].

A homoaromatic structure also best explains the n.m.r. characteristics of the anion **75** (see Table 15), which was derived from 1,6-methano-[10]annulene [219)]. This $10\pi9C$ system has peripheral protons which absorb from τ 3.19 to 5.28 (or *ca.* τ 2.1 to 4.2 on correction for charge density), and bridge protons which are about 2.5 ppm more shielded than the corresponding protons in *65*, the compound formed on protonation of the anion.

$$\begin{array}{c} CH_2 \\ | \\ SO \\ | \\ CH_3 \end{array}$$

65

1,3 overlap does not seem to be particularly significant in compounds where resonance stabilization of an incipient charge is not necessary (see footnote p. 200). Thus, neutral compounds which appear to have the option of homoaromaticity seem to exist as unconjugated polyolefins (**76, 77** and **78**, see Table 16) [220–223].

A potentially anti-homoaromatic species (**79**, see Table 16) has been generated by protonation of 1,6-methano-[10]annulene at the 2-position [224]. As expected, 1,3 overlap is not significant and the compound can be formulated as a classical carbonium ion. In fact the abnormally large H(2a) to H(2a) coupling constant of −24.9 Hz indicates that the 1,3 distance has been maximized in order to avoid the expected destabilization, and thus there is no evidence for a paramagnetic ring current in this system.

In spite of the fact that aromaticity is usually associated with (reasonable) planarity of the π-system, Goldstein [225] has set forth certain rules which, though based on HMO theory, apply to bicyclic hydrocarbons. Perhaps because the interactions postulated are expected to be fairly small, evidence for this type of stabilization is difficult to adduce from the observed n.m.r., and will have to await a more thorough-going investigation.

However, two compounds which are potentially within this definition have been prepared and their spectra obtained (**80** and **81**, see Table 16). The 7-norbornadienyl cation, **80** [226,227], definitely prefers not to avail itself of bicycloaromatic stabilization of the bridge carbonium ion by "both" double bonds, as the expected C_{2h} symmetry is ruled out by the n.m.r. spectrum which in fact resembles that of the 7-norbornenyl cation, **71**. Thus, bishomoconjugation only is observed, and the barrier to bridge flipping is estimated to be ⩾19.6 kcal/mole [228].

The required symmetry appears to be present in the case of the bicyclo[3.2.2]nonatrienyl anion, **81** [229], as the spectrum in diglyme is invariant down to −35 °C. Thus, if bridge flipping occurs in this system its free energy barrier must be ⩽11.8 kcal/mole. Once again the n.m.r. results are not definitive with respect to the degree of non-classical resonance in this compound.

Table 16*.

Compound	Chemical shifts (τ)	Coupling constants (Hz)	Solvent	Counterion	Conc. temp.	Ref.
76	6.05 (H_A) 7.74 (H_B) 4.44 (H_X)	$J_{A,B} = 12.4$	$CDCl_3$		—40 °C	220)
	7.04 (H_A, H_B) 4.53 (H_X)		$CDCl_3$		80 °C	
77	4.23 (olefinic protons) 7.76 (aliphatic protons)		$CDCl_3$		37 °C	222)
78	4.32 (olefinic protons) 6.23 (aliphatic protons)		$CDCl_3$			223)

* See footnote to Table 4, p. 134.

Table 16 (continued)

Compound	Chemical shifts (τ)	Coupling constants (Hz)	Solvent	Counterion	Conc. temp.	Ref.
79	5.54 $\}$ (H$_{2a}$, H$_{2b}$) 6.10 $\}$ 2.21 (H$_3$) 2.56 (H$_4$) 1.52 (H$_5$) 2.12 (H$_7$) 2.43 (H$_8$) 1.96 (H$_9$) 2.68 (H$_{10}$) 7.78 $\}$ (H$_{11a}$, 8.42 $\}$ H$_{11b}$)	$J_{2a,2b} = -24.9 \pm 0.1$ $J_{2a,3} = 5.8 \pm 0.2$ $J_{3,4} = 10.0 \pm 0.5$ $J_{4,5} = 6.5 \pm 0.2$ $J_{7,8} = 9.0 \pm 0.5$ $J_{8,9} = 9.0 \pm 0.5$ $J_{9,10} = 8.0 \pm 0.5$ $J_{11a,11b} = -9.3 \pm 0.3$	FSO$_3$H— SO$_2$ClF		*ca.* —85 °C	[224]
80	2.42 (H$_2$, H$_3$) 3.74 (H$_5$, H$_6$) 4.73 (H$_1$, H$_4$) 6.52 (H$_7$)	$J_{1,2} = 6.1$ $J_{1,3} = 1.5$ $J_{2,3} = 4.6$ $J_{1,4} = 0.5$ $J_{2,7} = 2.7$ $J_{1,6} = 1.8$ $J_{1,5} = 1.8$ $J_{1,7} = 2.8$ $J_{5,7} = 1.0$	SO$_2$	BF$_4^-$	*ca.* 0.9 M	[227]
81	7.71 (H$_1$, H$_5$) 6.95 (H$_2$, H$_4$) 4.76 (H$_3$) 5.02 (H$_6$ to H$_9$)	$J_{1,2} = 7.6$ $J_{1,3} \sim 1.0$ $J_{1,5} \sim 0.7$ $J_{1,6} = 1.8$ $J_{1,7} = 6.4$ $J_{2,3} = 7.6$ $J_{2,7} \sim 1.5$ $J_{6,7} \sim 7.5$	DME-d_{10}	K$^+$		[229]

G. Addendum

[12]Annulene

[12]annulene, **82**, was originally postulated as the common intermediate in the photochemical and thermal equilibria between *66, 67* and *68* [231]. Some recent careful work has resulted in it being trapped at − 80 °C and

66   *67*   *68*

identified by n.m.r. [232]. At this temperature the spectrum consists of two bands of equal intensity due to the protons on the *cis-* and *trans-* double bonds. The equivalence results from rapid rotation of the *trans-* double bonds in this structure *69*. This process has an extremely low

69

energy of activation and its rate only becomes comparable with the n.m.r. time scale in the temperature range − 140 °C to − 170 °C. By an iterative line shape analysis it was possible to derive the chemical shifts in the absence of exchange as well as the kinetic parameters [233].

As pointed out above (p. 144) the inner protons are subject to very unfavorable steric interactions which accounts for the relatively small paramagnetic shielding of these protons.

A hydrazino bridged [12]annulene, believed to have the structure *70*, has been reported [234]. The proton chemical shifts do not indicate the

τ 4.78 (4H)
τ 4.87 (4H)

70

208

existence of a ring current, but the detailed geometry of this system has not yet been established.

[14]Annulenes

Professor Sondheimer informs us that he and his colleagues no longer regard the two known forms of monodehydro-[14]annulene as conformational isomers but believe them to be geometric isomers [99].

They have reached similar conclusions concerning the two isomeric [14]annulenes. Isomer A (p. 146) definitely has the structure 9 as previously assigned, but Isomer B probably has the configuration 71.

71

Bridged [14]Annulenes

Boekelheide and his coworkers [235,236] have now reported data for the parent hydrocarbon, *trans*-15,16-dihydropyrene, 83. As expected, the bridge protons are very strongly shielded. They have also prepared a monoaza analogue, *trans*-1,3,15,16-tetramethyl-15,16-dihydro-2-azapyrene, 84, the chemical shifts of the internal methyl groups of which indicate a slight attenuation of the ring current [237].

Professor Boekelheide has kindly informed us that his group has now obtained *cis*-15,16-dimethyldihydropyrene, 85. In this compound there is a significant decrease in the shielding of the internal methyl groups, suggesting a more serious departure from planarity of the carbocyclic perimeter than in the *trans*-isomer [235].

Bridged [18]Annulenes

A *meso*-thiaporphin 72 has been synthesized and is non-aromatic [238].

72

A porphin analogue **86**, in which the two bridging NH groups are replaced by sulphur atoms, has been reported and, as expected, its n.m.r. spectrum indicates a ring current comparable with other compounds in this series [238].

The n.m.r. spectra of NN'-dimethylporphyrins, **87** and **88**, have been reported [239]. The N-methyl groups must, of course, lie out of the plane of the macrocyclic ring and they are strongly shielded.

Heteronins: 10π-Systems

Four compounds, **89** [240], *73* [241], *74* [242] and *75* [242] in the nine membered 10π heteroannulene series have now been reported; these compounds are isoelectronic with the cyclononatetraenyl anion, **60**, a planar aromatic system. Of the three heteronin systems (N, O, S) reported however, only 1-H azonin, **89**, appears to possess aromatic character, as judged by the n.m.r. spectra [240].

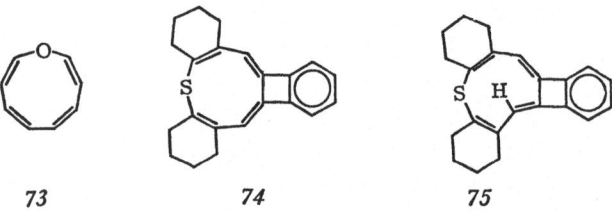

73   *74*   *75*

1,4-Bishomotropylium Ion

An n.m.r. study [243] of the bicyclo[4.3.0]nonatrienyl cation now provides the most clear-cut evidence for the existence of bishomoaromatic species. The 1,4-bishomotropylium ion [243], **90**, shows n.m.r. absorptions which are in general quite reminiscent of those observed for the monohomotropylium ion, **73**, and the tropylium ion, **59**.

Bridged [14]Annulenes

Professor Vogel [244] and his co-workers have now reported the synthesis of the aromatic compound 1,6;8,13-butano-[14]annulene, **91** [245]. In addition, an X-ray structural analysis of 1,6;8,13-propano-[14]annulene has appeared [246].

Table 17*. *Addendum*

Compound	Chemical shifts (τ)	Coupling constants (Hz)	Solvent, temperature	Ref.
82 [12]Annulene	Mobile spectrum 3.12 (H_1, H_2) 4.03 (H_3, H_4) Inferred non-mobile spectrum[a]) 1.98 ($q.$, H_1) 4.35 ($q.$, H_2) 3.82 ($q.$, H_3 or H_4) 4.39 ($q.$, H_4 or H_3)		THF-d_8 $-80\,°C$ THF-d_8/ CD_3OCD_3 (1:1 by vol.) $<-170\,°C$	232) 233)
83 *trans*-15,16-Dihydropyrene	1.42 ($s.$, H_4, H_5, H_9, H_{10}) 1.50 ($d.$, H_1, H_3, H_6, H_8) 1.98—2.11 ($m.$, H_2, H_7) 15.49 ($s.$, H_{15}, H_{16})	$J_{1,2} = 7.5$ $(=J_{2,3}=$ $J_{6,7} = J_{7,8})$	Cyclo-hexane	235, 236)
84 *trans*-1,3,15,16-Tetra-methyl-15,16-dihydro-2-azapyrene	1.0—1.9 (7H, $m.$, ring protons) 6.57 (6H, $s.$, external methyls) 13.75 (3H, $s.$) } (internal 13.80 (3H, $s.$) } methyls)		CDCl$_3$	237)

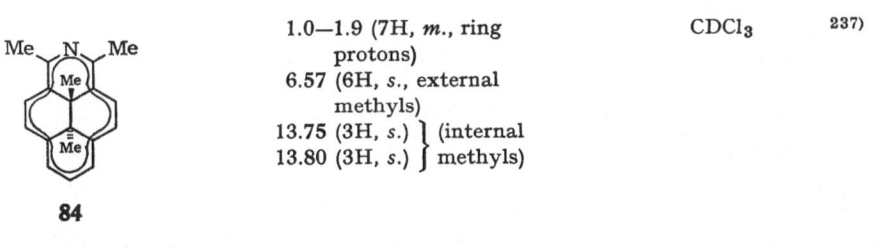

* See footnote to Table 4, p. 134.

Table 17 (continued)

Compound	Chemical shifts (τ)	Coupling constants (Hz)	Solvent, temperature	Ref.
85 cis-15,16-Dimethyl-dihydropyrene	1.26 (H_4, H_5, H_9, H_{10}) 1.76 ($d.$, H_1, H_3, H_6, H_8) 2.50 ($m.$, H_2, H_7) 12.06 (CH_3)	$J_{1,2} = 7.5$ $(= J_{2,3} = J_{6,7} = J_{7,8})$		235)
86	-0.71 ⎫ meso -0.68 ⎭ protons		$CDCl_3$	238)
87 R=Me NN′-Dimethylaetiopor-phyrin-I monohydro-chloride	-0.21 (1H) ⎫ meso -0.33 (3H) ⎭ protons 13.76 (1H, N—H) 15.88 (3H) ⎫ N—CH_3 15.92 (3H) ⎭		$CDCl_3$	239)
88 R=Et NN′-Dimethyloctaethyl-porphyrin monohydro-chloride	-0.43 (3H) ⎫ meso -0.10 (1H) ⎭ protons 13.70 (1H, N—H) 15.70 (6H, N—CH_3)		$CDCl_3$	239)

Table 17 (continued)

Compound	Chemical shifts (τ)	Coupling constants (Hz)	Solvent, temperature	Ref.
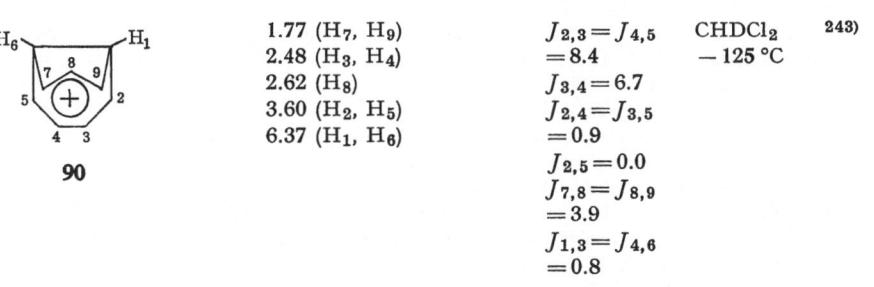 **89** 1-H Azonin	2.85 (2H, *d.d.*, H_2, H_9) 3.27 (4H, *m.*, H_4, H_5, H_6, H_7) 3.92—4.25 (2H, *m.*, H_3, H_8)		CDCl$_3$ Unchanged from -50 °C to 35 °C	240)
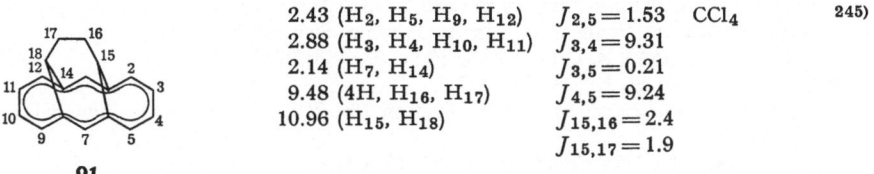 **90**	1.77 (H_7, H_9) 2.48 (H_3, H_4) 2.62 (H_8) 3.60 (H_2, H_5) 6.37 (H_1, H_6)	$J_{2,3} = J_{4,5}$ $= 8.4$ $J_{3,4} = 6.7$ $J_{2,4} = J_{3,5}$ $= 0.9$ $J_{2,5} = 0.0$ $J_{7,8} = J_{8,9}$ $= 3.9$ $J_{1,3} = J_{4,6}$ $= 0.8$	CHDCl$_2$ -125 °C	243)
91 1,6;8,13-Butano-[14]annulene	2.43 (H_2, H_5, H_9, H_{12}) 2.88 (H_3, H_4, H_{10}, H_{11}) 2.14 (H_7, H_{14}) 9.48 (4H, H_{16}, H_{17}) 10.96 (H_{15}, H_{18})	$J_{2,5} = 1.53$ $J_{3,4} = 9.31$ $J_{3,5} = 0.21$ $J_{4,5} = 9.24$ $J_{15,16} = 2.4$ $J_{15,17} = 1.9$	CCl$_4$	245)

a) Chemical shifts originally quoted relative to benzene. The conversion to the τ scale is made by adding 2.73 ppm to the quoted values.

H. References

1) a) Hückel, E.: Z. Physik *70*, 204 (1931);
   b) — Grundzüge der Theorie ungesättigter und aromatischer Verbindungen, pp. 71—85. Berlin: Verlag Chemie 1938.
2) a) Dewar, M. J. S.: Rev. Mod. Phys. *35*, 586 (1963);
   b) Chung, A. L. H., Dewar, M. J. S: J. Chem. Phys. *42*, 756 (1965);
   c) Dewar, M. J. S., Gleicher, G. J.: J. Am. Chem. Soc. *87*, 685 (1965);
   d) — — J. Am. Chem. Soc. *87*, 692 (1965);
   e) Ref. 3), and references therein.
3) Dewar, M. J. S.: Aromaticity, Special Publication No. 21, p. 177. London: Chemical Society 1967.
4) For a discussion see Ref. 5), Chap. 3.
5) Salem, L.: The Molecular Orbital Theory of Conjugated Systems. New York—Amsterdam: W. A. Benjamin, Inc. 1966.
6) For a discussion see Ref. 5), Chap. 8.
7) a) Longuet-Higgins, H. C., Salem, L.: Proc. Roy. Soc. (London) *A 251*, 172 (1959).
   b) — — Proc. Roy. Soc. (London) *A 257*, 445 (1960).
8) a) Kuhn, H.: J. Chem. Phys. *16*, 840 (1948);
   b) — J. Chem. Phys. *17*, 1198 (1949).
9) Dewar, M. J. S.: J. Chem. Soc. 3544 (1952).
10) Hultgren, G.O.: Ph.D.Thesis.California Institute of Technology. Pasadena (1966).
11) Vogel, E., Roth, H. D.: Angew. Chem. *76*, 145 (1964); Angew. Chem. Intern. Ed. Engl. *3*, 228 (1964).
12) For a general review see Jones, A. J.: Rev. Pure Appl. Chem. *18*, 253 (1968).
13) Clar, E.: Polycyclic Hydrocarbons. New York: Academic Press 1964.
14) Goeppert-Mayer, M., Sklar, A. L.: J. Chem. Phys. *6*, 645 (1938).
15) a) Pople, J. A.: Trans. Faraday Soc. *49*, 1375 (1953).
    b) Brickstock, A., Pople, J. A.: Trans. Faraday Soc. *50*, 901 (1954);
    c) Pariser, R., Parr, R. G.: J. Chem. Phys. *21*, 466, 767 (1953).
16) For a discussion see Ref. 5), Chap. 7.
17) Vogel, E.: Aromaticity, Special Publication No. 21, p. 113. London: Chemical Society 1967.
18) Traetteberg, M.: Acta Chem. Scand. *20*, 1724 (1966).
19) Haberditzl, W.: Sitzber. Deut. Akad. Wiss. Berlin, Kl. Chem. Geol. Biol. No. *2* (1964); Angew. Chem. Intern. Ed. Engl. *5*, 288 (1966).
20) Pacault, A., Hoarau, J., Marchand, A.: Advan. Chem. Phys. *3*, 171 (1961).
21) For a general review see: Bothner-By, A. A., Pople, J. A.: Ann. Rev. Phys. Chem. *16*, 43 (1965).
22) Dorfman, G.: Diamagnetism and the Chemical Bond. New York: Edward Arnold. London 1965.
23) Van Vleck, J. H.: The Theory of Electric and Magnetic Susceptibilities. Oxford: University Press 1932.

24) See Ref. 25), and references therein.
25) Pauling, L.: J. Chem. Phys. *4*, 673 (1936).
26) See also: Lonsdale, K.: Proc. Roy. Soc. (London) *159 A*, 149 (1937).
27) a) London, F.: J. Chem. Phys. *5*, 837 (1937).
    b) — J. Phys. Radium *8*, 397 (1937).
28) a) Dauben, H. J., Wilson, J. D., Laity, J. L.: J. Am. Chem. Soc. *90*, 811 (1968).
    b) — — — J. Am. Chem. Soc. *91*, 1991 (1969).
29) Pople, J. A., Untch, K. G.: J. Am. Chem. Soc. *88*, 4811 (1966).
30) Baer, F., Kuhn, H., Regel, W.: Z. Naturforsch. *22a*, 103 (1967).
31) a) Longuet-Higgins, H. C.: Aromaticity, Special Publication No. 21, p. 109. London: Chemical Society 1967.
    b) Nakajima, T., Kohda, S.: Bull. Soc. Chem. Japan *39*, 804 (1966).
32) For a discussion see Ref. 5), Chap. 4.
33) Maddox, I. J., McWeeny, R.: J. Chem. Phys. *36*, 2353 (1962).
34) Dailey, B. P.: J. Chem. Phys. *41*, 2304 (1964).
35) McWeeny, R.: Proc. Phys. Soc. (London) *64A*, 261 (1951).
36) a) Pople, J. A.: J. Chem. Phys. *37*, 53 (1962);
    b) Hameka, H. F.: J. Chem. Phys. *37*, 3008 (1962);
    c) Pople, J. A.: J. Chem. Phys. *37*, 3009 (1962);
    d) — Discussions Faraday Soc. *34*, 67 (1962).
37) Pullman, B., Pullman, A.: Les Théories électroniques de la chimie organique, Chap. IX. Paris: Masson 1952 (written in collaboration with G. Berthier).
38) See: Pople, J. A.: Mol. Phys. *1*, 175 (1958).
39) For a discussion see Ref. 33).
40) Hoarau, J.: Ann. Chim. (Paris) *1*, 560 (1956).
41) Davies, D. W.: Trans. Faraday Soc. *57*, 2081 (1961).
42) a) Pople, J. A.: J. Chem. Phys. *41*, 2559 (1964);
    b) Ferguson, A. F., Pople, J. A.: J. Chem. Phys. *42*, 1560 (1965).
43) Baudet, J., Guy, J., Tillieu, J.: J. Phys. Radium *21*, 600 (1960).
44) a) Hoarau, J., Lumbroso, N., Pacault, A.: Compt. Rend. *242*, 1702 (1956);
    b) Bailey, N. A., Gerloch, M., Mason, R.: Mol. Phys. *10*, 327 (1966).
45) For a discussion see Ref. 5), pp. 194—205.
46) Berthier, G., Mayot, M., Pullman, B.: J. Phys. Radium *12*, 717 (1951).
47) Craig, D. P.: Non-benzenoid Aromatic Compounds; ed. D. Ginsburg. New York: Interscience Publishers 1959.
48) Emsley, J. W., Feeney, J., Sutcliffe, L. H.: High Resolution Nuclear Magnetic Resonance Spectroscopy, Vol. I and II. Oxford: Pergamon 1965.
49) Jackman, L. M., Sternhell, S.: Applications of Nuclear Magnetic Resonance Spectroscopy in Organic Chemistry, 2nd edition. Oxford: Pergamon 1969.
50) Waugh, J. S., Fessenden, R. W.: J. Am. Chem. Soc. *79*, 846 (1957); J. Am. Chem. Soc. *80*, 6697 (1958).
51) Ref. 5), p. 209.
52) Pople, J. A.: J. Chem. Phys. *24*, 1111 (1956).
53) Johnson, C. E., Bovey, F. A.: J. Chem. Phys. *29*, 1012 (1958).
54) Abraham, R. J.: Mol. Phys. *4*, 145 (1961).
55) McWeeny, R.: Mol. Phys. *1*, 311 (1958).
56) Hall, G. G., Hardisson, A.: Proc. Roy. Soc. (London) *A 268*, 328 (1962).
57) Black, P. J., Brown, R. D., Heffernan, M. L.: Australian J. Chem. *20*, 1305, 1325 (1967).
58) Hall, G. G., Hardisson, A., Jackman, L. M.: Tetrahedron *19* Suppl. 2, 101 (1963).
59) Ref. 5), pp. 210—214.
60) Haddon, R. C.: to be published.

References

61) a) Musher, J. I.: J. Chem. Phys. *43*, 4081 (1965);
   b) — Advan. Magnetic Resonance *2*, 177 (1966).
62) Gaidis, J. M., West, R.: J. Chem. Phys. *46*, 1218 (1967).
63) Musher, J. I.: J. Chem. Phys. *46*, 1219 (1967).
64) Amos, T., Musher, J. I.: J. Chem. Phys. *49*, 2158 (1968).
65) Cooper, M. A., Manatt, S. L.: J. Am. Chem. Soc. *91*, 6325 (1969).
66) Sondheimer, F., Calder, I. C., Elix, J. A., Gaoni, Y., Garratt, P. J., Grohmann, K., di Maio, G., Mayer, J., Sargent, M. V., Wolovsky, R.: Aromaticity, Special Publication No. 21, p. 75. London: Chemical Society 1967.
67) Emerson, G. F., Watts, L., Pettit, R.: J. Am. Chem. Soc. *87*, 131 (1965).
68) Bovey, F. A.: N.M.R. Data Tables for Organic Compounds, Vol. 1. New York: Interscience Publishers 1967.
69) Anet, F. A. L., Bourn, A. J. R., Lin, Y. S.: J. Am. Chem. Soc. *86*, 3576 (1964).
70) Dewar, M. J. S., Harget, A., Haselbach, E.: J. Am. Chem. Soc. *91*, 7521 (1969).
71) Mislow, K.: J. Chem. Phys. *20*, 1489 (1952).
72) Masamune, S., Seidner, R. T.: Chem. Commun. 542 (1969).
73) Mitchell, R. H., Sondheimer, F.: J. Am. Chem. Soc. *90*, 530 (1968).
74) Grohmann, K., Sondheimer, F.: J. Am. Chem. Soc. *89*, 7119 (1967).
75) Bindra, A. P., Elix, J. A., Sargent, M. V.: Tetrahedron Letters 4335 (1968).
76) Masamune, S., Chin, C. G., Hojo, K., Seidner, R. T.: J. Am. Chem. Soc. *89*, 4804 (1967).
77) van Tamelen, E. E., Burkoth, T. L.: J. Am. Chem. Soc. *89*, 151 (1967).
78) Untch, K. G., Wysocki, D. C.: J. Am. Chem. Soc. *88*, 2608 (1966).
79) Sondheimer, F., Wolovsky, R., Garratt, P. J., Calder, I. C.: J. Am. Chem. Soc. *88*, 2610 (1966).
80) Okamura, W. H., Sondheimer, F.: J. Am. Chem. Soc. *89*, 5991 (1967).
81) Wolovsky, R., Sondheimer, F.: J. Am. Chem. Soc. *87*, 5720 (1965).
82) Untch, K. G., Wysocki, D. C.: J. Am. Chem. Soc. *89*, 6386 (1967).
83) Wilke, G.: Angew. Chem. *69*, 397 (1957).
84) Cope, A. C., Fenton, S. W.: J. Am. Chem. Soc. *73*, 1195 (1951), and first four references therein.
85) Mayer, J., Sondheimer, F.: J. Am. Chem. Soc. *88*, 602 (1966).
86) Sondheimer, F., Gaoni, Y.: J. Am. Chem. Soc. *82*, 5765 (1960).
87) — —, Jackman, L. M., Bailey, N. A., Mason, R.: J. Am. Chem. Soc. *84*, 4595 (1962).
88) Bailey, N. A., Mason, R.: Proc. Roy. Soc. (London) *A 290*, 94 (1966).
89) Gaoni, Y., Sondheimer, F.: J. Am. Chem. Soc. *86*, 521 (1964).
90) Sondheimer, F.: Proc. Roy. Soc. (London) *A 297*, 173 (1967).
91) Jackman, L. M., Sondheimer, F., Amiel, Y., Ben-Efraim, D. A., Gaoni, Y., Wolovsky, R., Bothner-By, A. A.: J. Am. Chem. Soc. *84*, 4307 (1962).
92) Sondheimer, F.: Pure Appl. Chem. *7*, 363 (1963).
93) Gaoni, Y., Sondheimer, F.: Proc. Chem. Soc. (London) 299 (1964).
94) Bregman, J.: Nature *194*, 679 (1962).
95) Gaoni, Y., Melera, A., Sondheimer, F., Wolovsky, R.: Proc. Chem. Soc. (London) 397 (1964).
96) Calder, I. C., Gaoni, Y., Sondheimer, F.: J. Am. Chem. Soc. *90*, 4946 (1968).
97) Sondheimer, F., Gaoni, Y.: J. Am. Chem. Soc. *83*, 4863 (1961).
98) Calder, I. C., Gaoni, Y., Garratt, P. J., Sondheimer, F.: J. Am. Chem. Soc. *90*, 4954 (1968).
99) Sondheimer, F.: private communication.
100) Garratt, P. J.: private communication.

[101] Calder, I. C., Garratt, P. J., Sondheimer, F.: Chem. Commun. 41 (1967).
[102] a) Forsén, S., Hoffman, R. A.: Acta. Chem. Scand. 17, 1787 (1963);
 b) — — J. Chem. Phys. 39, 2892 (1963);
 c) — — J. Chem. Phys. 40, 1189 (1964).
[103] Schröder, G., Oth, J. F. M.: Tetrahedron Letters 4083 (1966).
[104] Oth, J. F. M., Gilles, J-M.: Tetrahedron Letters 6259 (1968).
[105] a) Johnson, S. M., Paul, I. C.: J. Am. Chem. Soc. 90, 6555 (1968).
 b) — —, King, G. S. D.: J. Chem. Soc. B 643 (1970).
[106] Schröder, G., Kirsch, G., Oth, J. F. M.: Tetrahedron Letters 4575 (1969).
[107] Wolovsky, R.: J. Am. Chem. Soc. 87, 3638 (1965).
[108] Ojima, J., Katakami, T., Nakaminami, G., Nakagawa, M.: Tetrahedron letters 1115 (1968).
[109] Bregman, J.: see footnote 5, Ref. [95].
[110] Sondheimer, F., Wolovsky, R.: J. Am. Chem. Soc. 84, 260 (1962).
[111] Calder, I. C., Garratt, P. J., Longuet-Higgins, H. C., Sondheimer, F., Wolovsky, R.: J. Chem. Soc. C 1041 (1967).
[112] Sondheimer, F., Wolovsky, R., Amiel, Y.: J. Am. Chem. Soc. 84, 274 (1962).
[113] Hirshfeld, F. L., Rabinovich, D.: Acta Cryst. 19, 235 (1965).
[114] Bregman, J., Hirshfeld, F. L., Rabinovich, D., Schmidt, G. M. J.: Acta Cryst. 19, 227 (1965).
[115] Sondheimer, F., Gaoni, Y.: J. Am. Chem. Soc. 83, 1259 (1961).
[116] — — J. Am. Chem. Soc. 84, 3520 (1962).
[117] Calder, I. C., Sondheimer, F.: Chem. Commun. 904 (1966).
[118] Leznoff, C. C., Sondheimer, F.: J. Am. Chem. Soc. 89, 4247 (1967).
[119] Castellano, S., Sun, C.: J. Am. Chem. Soc. 88, 4741 (1966).
[120] Vogel, E., Böll, W. A.: Angew. Chem. 76, 784 (1964); Angew. Chem. Intern. Ed. Engl. 3, 642 (1964).
[121] Günther, H.: Z. Naturforsch. 20b, 948 (1965).
[122] Vogel, E., Maier, W., Eimer, J.: Tetrahedron Letters 655 (1966).
[123] Bloomfield, J. J., Irelan, J. R. S.: Tetrahedron Letters 2971 (1966).
[124] Böll, W. A.: Tetrahedron Letters 2595 (1968).
[125] Vogel, E., Böll, W. A., Biskup, M.: Tetrahedron Letters 1569 (1966).
[126] —, Schröck, W., Böll, W. A.: Angew. Chem. 78, 753 (1966); Angew. Chem. Intern. Ed. Engl. 5, 732 (1966).
[127] — Chimia 22, 21 (1968).
[128] Fischer, E. O., Rühle, H., Vogel, E., Grimme, W.: Angew. Chem. 78, 548 (1966); Angew. Chem. Intern. Ed. Engl. 5, 518 (1966).
[129] Dobler, M., Dunitz, J. D.: Helv. Chim. Acta. 48, 1429 (1965).
[130] Vogel, E., Schröck, W.: unpublished results, see Ref. [140]; footnote (92) therein.
[131] Gerson, F., Heilbronner, E., Böll, W. A., Vogel, E.: Helv. Chim. Acta. 48, 1494 (1965).
[132] Vogel, E., Weyres, F., Lepper, H., Rautenstrauch, V.: Angew. Chem. 78, 754 (1966); Angew. Chem. Intern. Ed. Engl. 5, 732 (1966).
[133] — Grimme, W., Korte, S.: Tetrahedron Letters 3625 (1965)
[134] Rautenstrauch, V., Scholl, H.-J., Vogel, E.: Angew. Chem. 80, 278 (1968); Angew. Chem. Intern. Ed. Engl. 7, 288 (1968).
[135] Günther, H., Hinrichs, H.-H.: Tetrahedron 24, 7033 (1968).
[136] Vogel, E.: private communication.
[137] Shani, A., Sondheimer, F.: J. Am. Chem. Soc. 89, 6310 (1967).
[138] Sondheimer, F., Shani, A.: J. Am. Chem. Soc. 86, 3168 (1964).
[139] Vogel, E., Biskup, M., Pretzer, W., Böll, W. A.: Angew. Chem. 76, 785 (1964); Angew. Chem. Intern. Ed. Engl. 3, 642 (1964).

References

140) Vogel, E., Günther, H.: Angew. Chem. *79*, 429 (1967); Angew. Chem. Intern. Ed. Engl. *6*, 385 (1967).
141) Bailey, N. A., Mason, R.: Chem. Commun. 1039 (1967).
142) Biskup, M.: Dissertation, Köln (1966).
143) Vogel, E., Pretzer, W., Böll, W. A.: Tetrahedron Letters 3613 (1965).
144) Blattmann, H.-R., Böll, W. A., Heilbronner, E., Hohlneicher, G., Vogel, E., Weber, J.-P.: Helv. Chim. Acta. *49*, 2017 (1966).
145) Windgassen, R. J., jr., Saunders, W. H., jr., Boekelheide, V.: J. Am. Chem. Soc. *81*, 1459 (1959).
146) Galbraith, A., Small, T., Barnes, R. A., Boekelheide, V.: J. Am. Chem. Soc. *83*, 453 (1961).
147) Hardisson, A.: Ph. D. Thesis, Imperial College, London (1962).
148) Boekelheide, V., Gerson, F., Heilbronner, E., Meuche, D.: Helv. Chim. Acta, *46*, 1951 (1963).
149) Jackman, L. M., Porter, Q. N., Underwood, G. R.: Australian J. Chem. *18*, 1221 (1965).
150) Farquhar, D., Leaver, D.: Chem. Commun. 24 (1969).
151) Boekelheide, V., Phillips, J. B.: Proc. Natl. Acad. Sci. U.S. *51*, 550 (1964).
152) — — J. Am. Chem. Soc. *89*, 1695 (1967).
153) Hanson, A. W.: Acta Cryst. *18*, 599 (1965).
154) Boekelheide, V., Phillips, J. B.: J. Am. Chem. Soc. *85*, 1545 (1963).
155) Phillips, J. B., Molyneux, R. J., Sturm, E., Boekelheide, V.: J. Am. Chem. Soc. *89*, 1704 (1967).
156) Boekelheide, V., Miyasaka, T.: J. Am. Chem. Soc. *89*, 1709 (1967).
157) Mitchell, R. H., Klopfenstein, C. E., Boekelheide, V.: J. Am. Chem. Soc. *91*, 4931 (1969).
158) Hanson, A. W.: Acta Cryst. *23*, 476 (1967).
159) Haberland, U.: Dissertation, Köln (1968).
160) Vogel, E., Biskup, M., Vogel, A., Günther, H.: Angew. Chem. *78*, 755 (1966); Angew. Chem. Intern. Ed. Engl. *5*, 734 (1966).
161) Ganis, P., Dunitz, J. D.: Helv. Chim. Acta, *50*, 2369 (1967).
162) Bremser, W.: Dissertation, Köln (1968).
163) Winstein, S., Carter, P., Anet, F. A. L., Bourn, A. J. R.: J. Am. Chem. Soc. *87*, 5247 (1965).
164) Vogel, E., Vogel, A., Kübbeler, H.-K., Sturm, W.: Angew. Chem. *82*, 512 (1970); Angew. Chem. Intern. Ed. Engl. *9*, 514 (1970).
165) Bremser, W., Roberts, J. D., Vogel, E.: Tetrahedron Letters 4307 (1969).
166) Becker, E. D., Bradley, R. B.: J. Chem. Phys. *31*, 1413 (1959).
167) a) Ellis, J., Jackson, A. H., Kenner, G. W., Lee, J.: Tetrahedron Letters No. 2, 23 (1960).
   b) Becker, E. D., Bradley, R. B., Watson, C. J.: J. Am. Chem. Soc. *83*, 3743 (1961).
   c) Caughey, W. S., Koski, W. S.: Biochemistry *1*, 923 (1962).
168) Abraham, R. J., Jackson, A. H., Kenner, G. W.: J. Chem. Soc. 3468 (1961).
169) — Mol. Phys. *4*, 145 (1961).
170) Kowalsky, A., Cohn, M.: Ann. Rev. Biochem. *33*, 481 (1964).
171) Bonnett, R., Gale, I. A. D., Stephenson, G. F.: J. Chem. Soc. C 1168 (1967).
172) Abraham, R. J., Burbidge, P. A., Jackson, A. H., Kenner, G. W.: Proc. Chem. Soc. (London) 134 (1963).
173) Bonnett, R., Stephenson, G. F.: J. Org. Chem. *30*, 2791 (1965).
174) Pople, J. A., Schneider, W. G., Bernstein, H. J.: High Resolution Nuclear Magnetic Resonance, p. 407. New York: McGraw-Hill 1959.

175) Gil, V. M. S., Murrell, J. N.: Trans. Faraday Soc. *60*, 248 (1964).

176) Broadhurst, M. J., Grigg, R., Johnson, A. W.: Chem. Commun. 1480 (1969).

177) Badger, G. M., Elix, J. A., Lewis, G. E.: Proc. Chem. Soc. (London) 82 (1964).

178) — — — Australian J. Chem. *18*, 70 (1965).

179) Coulson, C. A., Poole, M. D.: Proc. Chem. Soc. (London) 220 (1964).

180) Badger, G. M., Elix, J. A., Lewis, G. E., Singh, U. P., Spotswood, T. M.: Chem. Commun. 269 (1965).

181) — Lewis, G. E., Singh, U. P.: Australian J. Chem. *19*, 257 (1966).

182) — — — Australian J. Chem. *20*, 1635 (1967).

183) — — — Spotswood, T. M.: Chem. Commun. 492 (1965).

184) — — — Australian J. Chem. *19*, 1461 (1966).

185) — Elix, J. A., Lewis, G. E.: Australian J. Chem. *19*, 1221 (1966).

186) Elix, J. A.: Chem. Commun. 343 (1968).

187) — Sargent, M. V.: J. Am. Chem. Soc. *90*, 1631 (1968).

188) Broadhurst, M. J., Grigg, R., Johnson, A. W.: Chem. Commun. 23 (1969); see also corrigenda, Chem. Commun. 1080 (1969).

189) Elix, J. A.: Australian J. Chem. *22*, 1951 (1969).

190) Schaefer, T., Schneider, W. G.: Can. J. Chem. *41*, 966 (1963).

191) Dailey, B. P., Gawer, A., Neikam, W. C.: Discussions Faraday Soc. *34*, 18 (1962).

192) Spiesecke, H., Schneider, W. G.: Tetrahedron Letters 468 (1961).

193) Ref. 49), p. 68.

194) Ref. 48) p. 781.

195) Breslow, R., Groves, J. T., Ryan, G.: J. Am. Chem. Soc. *89*, 5048 (1967).

196) Farnum, D. G., Mehta, G., Silberman, R. G.: J. Am. Chem. Soc. *89*, 5048 (1967).

197) Ref. 206), Footnote 18.

198) Fraenkel, G., Carter, R. E., McLachlan, A., Richards, J. H.: J. Am. Chem. Soc. *82*, 5846 (1960).

199) Leto, J. R., Cotton, F. A., Waugh, J. S.: Nature *180*, 978 (1957).

200) Ref. 5) p. 193.

201) Katz, T. J., Garratt, P. J.: J. Am. Chem. Soc. *85*, 2852 (1963).

202) LaLancette, E. A., Benson, R. E.: J. Am. Chem. Soc. *85*, 2853 (1963).

203) Grimme, W., Hoffman, H., Vogel, E.: Angew. Chem. *77*, 348 (1965); Angew. Chem. Intern. Ed. Engl. *4*, 354 (1965).

204) a) — Kaufhold, M., Dettmeier, U., Vogel, E.: Angew. Chem. *78*, 643 (1966);
b) Radlick, P., Rosen, W.: J. Am. Chem. Soc. *88*, 3461 (1966). Angew. Chem. Intern. Ed. Engl. *5*, 604 (1966).

205) Griffiths, J., Sondheimer, F.: J. Am. Chem. Soc. *91*, 7518 (1969).

206) Olah, G. A., Bollinger, J. M., White, A. M.: J. Am. Chem. Soc. *91*, 3667 (1969).

207) Deno, N. C., Richey, H. G., jr., Friedman, N., Hodge, J. D., Houser, J. J., Pittman, C. U., jr.: J. Am. Chem. Soc. *85*, 2991 (1963).

208) Katz, T. J.: J. Am. Chem. Soc. *82*, 3785 (1960).

209) von Fritz, H. P., Keller, H.: Z. Naturforsch. *16b*, 231 (1961).

210) Oth, J. F. M., Anthoine, G., Gilles, J.-M.: Tetrahedron Letters 6265 (1968).

211) Katz, T. J., Gold, E. H.: J. Am. Chem. Soc. *86*, 1600 (1964).

212) Winstein, S.: Aromaticity, Special Publication No. 21, p. 5. London: Chemical Society 1967.

213) Brookhart, M., Diaz, A., Winstein, S.: J. Am. Chem. Soc. *88*, 3135 (1966).

214) Richey, H. G., jr., Lustgarten, R. K.: J. Am. Chem. Soc. *88*, 3136 (1966).

215) Winstein, S., Ogliaruso, M., Sakai, M., Nicholson, J. M.: J. Am. Chem. Soc. *89*, 3656 (1967).

216) von Rosenberg, J. L., jr., Mahler, J. E., Pettit, R.: J. Am. Chem. Soc. *84*, 2842 (1962).

References

217) Winstein, S., Kaesz, H. D., Kreiter, C. G., Friedrich, E. C.: J. Am. Chem. Soc. 87, 3267 (1965).
218) Ogliaruso, M., Rieke, R., Winstein, S.: J. Am. Chem. Soc. 88, 4731 (1966).
219) Böll, W. A.: Tetrahedron Letters 5531 (1968).
220) Untch, K. G., Kurland, R. J.: J. Mol. Spectry. 14, 156 (1964).
221) a) Radlick, P., Winstein, S.: J. Am. Chem. Soc. 85, 344 (1963);
   b) Untch, K. G.: J. Am. Chem. Soc. 85, 345 (1963);
   c) — Kurland, R. J.: J. Am. Chem. Soc. 85, 346 (1963).
   d) Roth, W. R.: Ann. 671, 10 (1964);
   e) — Bang, W. B., Goebel, P., Sass, R. L., Turner, R. B., Yü, A. P.: J. Am. Chem. Soc. 86, 3178 (1964).
   f) Winstein, S., Lossing, F. P.: J. Am. Chem. Soc. 86, 4485 (1964).
222) Untch, K. G., Martin, D. J.: J. Am. Chem. Soc. 87, 3518 (1965).
223) Woodward, R. B., Fukunaga, T., Kelly, R. C.: J. Am. Chem. Soc. 86, 3162 (1964).
224) Warner, P., Winstein, S.: J. Am. Chem. Soc. 91, 7785 (1969).
225) Goldstein, M. J.: J. Am. Chem. Soc. 89, 6357 (1967).
226) Story, P. R., Saunders, M.: J. Am. Chem. Soc. 84, 4876 (1962).
227) — Snyder, L. C., Douglass, D. C., Anderson, E. W., Kornegay, R. L.: J. Am. Chem. Soc. 85, 3630 (1963).
228) Brookhart, M., Lustgarten, R. K., Winstein, S.: J. Am. Chem. Soc. 89, 6352 (1967).
229) Grutzner, J. B., Winstein, S.: J. Am. Chem. Soc. 90, 6562 (1968).
230) Boekelheide, V., Hylton, T. A.: J. Am. Soc. 92, 3669 (1970).
231) Röttele, H., Martin, W., Oth, J. F. M., Schröder, G.: Chem. Ber. 102, 3985 (1969).
232) Oth, J. F. M., Röttele, H., Schröder, G.: Tetrahedron Letters 61 (1970).
233) — Gilles, J.-M., Schröder, G.: Tetrahedron Letters 67 (1970).
234) Paudler, W. W., Stephan, E. A.: J. Am. Chem. Soc. 92, 4468 (1970).
235) Boekelheide, V.: private communication.
236) Mitchell, R. H., Boekelheide, V.: J. Am. Chem. Soc. 92, 3510 (1970).
237) Boeckelheide, V., Pepperdine, W.: J. Am. Chem. Soc. 92, 3684 (1970).
238) Broadhurst, M. J., Grigg, R., Johnson, A. W.: Chem. Commun. 807 (1970).
239) Dearden, G. R., Jackson, A. H.: Chem. Commun. 205 (1970).
240) Anastassiou, A. G., Gebrian, J. H.: Tetrahedron Letters 825 (1970).
241) — Cellura, R. P.: Chem. Commun. 903 (1969).
242) Garratt, P. J., Holmes, A. B., Sondheimer, F., Vollhardt, K. P. C.: J. Am. Chem. Soc. 92, 4492 (1970).
243) Ahlberg, P., Harris, D. L., Winstein, S.: J. Am. Chem. Soc. 92, 4454 (1970).
244) a) Vogel, E.: Proc. Robert A. Welch Foundation Conferences on Chemical Research, XII. Organic Synthesis, Houston, Texas, 215 (1968);
   b) — Pure Appl. Chem. 20, 237 (1969).
245) — Sturm, W., Cremer, H.-D.: Angew. Chem. 82, 513 (1970); Angew. Chem. Intern. Ed. Engl. 9, 516 (1970).
246) Casalone, G., Gavezzotti, A., Mugnoli, A., Simonetta, M.: Angew. Chem. 82, 516 (1970); Angew. Chem. Intern. Ed. Engl. 9, 519 (1970).
247) Vogel, E., Haberland, U., Günther, H.: Angew. Chem. 82, 510 (1970); Angew. Chem. Intern. Ed. Engl. 9, 513 (1970).
248) — — Ick, J.: Angew. Chem. 82, 514 (1970); Angew. Chem. Intern. Ed. Engl. 9, 517 (1970).

Received March 31, 1970

NMR

Basic Principles and Progress
Grundlagen und Fortschritte

Editors: **P. Diehl, E. Fluck, R. Kosfeld**

Volume 1:

**P. Diehl and
C. L. Khetrapal:
NMR Studies of
Molecules Oriented
in the Nematic Phase
of Liquid Crystals**

**R. G. Jones:
The Use of Symmetry
in Nuclear Magnetic
Resonance**

53 figures. V, 174 pages. 1969
Cloth DM 39,—; US $ 10.80

Volume 2:

**H. J. Keller:
NMR-Untersuchungen an
Komplexverbindungen**

22 Abbildungen. III, 88 Seiten. 1970
Gebunden DM 32,—; US $ 8.80

Volume 3:

**O. Kanert and
M. Mehring:
Static Quadrupole Effects
in Disordered Cubic Solids**

**F. Noack:
Nuclear Magnetic
Relaxation Spectroscopy**

In preparation
Due Summer 1971.

Volume 4:

**Natural and Synthetic
High Polymers**

Lectures presented at the 7th Collo-
quium on NMR Spectroscopy held
in the Institut für Physikalische
Chemie, April 13-17, 1970 as part of
the 100th anniversary celebrations
of the Rheinisch-Westfälische
Technische Hochschule Aachen
Approx. 202 figures
Approx. 320 pages. 1971
Cloth DM 64,—; US $ 17.60

SPRINGER-VERLAG
BERLIN·HEIDELBERG·NEW YORK

In kritischen Übersichten werden in dieser Reihe Stand und Entwicklung aktueller chemischer Forschungsgebiete beschrieben. Sie wendet sich an alle Chemiker in Forschung und Industrie, die am Fortschritt ihrer Wissenschaft teilhaben wollen.

In der Regel werden nur Beiträge veröffentlicht, die ausdrücklich angefordert worden sind. Schriftleitung und Herausgeber sind aber für ergänzende Anregungen und Hinweise jederzeit dankbar. Manuskripte können in den „Fortschritten der chemischen Forschung" in Deutsch oder Englisch veröffentlicht werden.

Jedes Heft der Reihe ist auch einzeln käuflich.

This series presents critical reviews of the present position and future trends in modern chemical research. It is addressed to all research and industrial chemists who wish to keep abreast of advances in their subject.

As a rule, contributions are specially commissioned. The editors and publishers will, however, always be pleased to receive suggestions and supplementary information. Papers are accepted for "Topics in Current Chemistry" in either German or English.

Single issues may be purchased separately.